T0297096

Developments in Mathematics

VOLUME 35

Series Editors:
Krishnaswami Alladi, *University of Florida, Gainesville, FL, USA*
Hershel M. Farkas, *Hebrew University of Jerusalem, Jerusalem, Israel*

More information about this series at http://www.springer.com/series/5834

Christian Constanda • Dale Doty • William Hamill

Boundary Integral Equation Methods and Numerical Solutions

Thin Plates on an Elastic Foundation

 Springer

Christian Constanda
The Charles W. Oliphant Professor
 of Mathematical Sciences
Department of Mathematics
The University of Tulsa
Tulsa, Oklahoma, USA

Dale Doty
Department of Mathematics
The University of Tulsa
Tulsa, Oklahoma, USA

William Hamill
Department of Mathematics
The University of Tulsa
Tulsa, Oklahoma, USA

ISSN 1389-2177   ISSN 2197-795X   (electronic)
Developments in Mathematics
ISBN 978-3-319-26307-6   ISBN 978-3-319-26309-0   (eBook)
DOI 10.1007/978-3-319-26309-0

Library of Congress Control Number: 2016930553

Springer Cham Heidelberg New York Dordrecht London

Printed on acid-free paper

Springer International Publishing AG Switzerland is part of Springer Science+Business Media (www.springer.com)

For Lia, Jennifer, and Kathy

Preface

Many problems in mathematical physics and engineering are modeled by elliptic systems of partial differential equations. Well-known examples in this context are higher-dimensional steady-state heat conduction [11], acoustics (see [8, 15]), gravitational potential [10], fluid mechanics [14], and elasticity theory (plane strain, bending of thin plates, and stationary flexural oscillations—see [11, 12]). Among the solution techniques for such problems, a prominent role is played by boundary integral equation methods (BIEMs), which, apart from being powerful and elegant, have some decisive advantages over other procedures, an important one being that they change a problem from its formulation in terms of an unbounded differential operator to one for an integral operator, thus making it more appealing and tractable from an analytic viewpoint. In essence, the ellipticity of the problems is shifted to the boundary, where it gives rise to integral equations that are then solved in suitable function spaces [6]. BIEMs also yield closed form solutions, which renders them very useful for numerical computation.

The many different types of BIEMs developed so far [6] can be divided into two main categories: direct and indirect. The former employ what might be called 'designer' solutions, of a form chosen purely for mathematical convenience. By contrast, the unknown functions in the latter have physical significance; in elasticity, for example, they may characterize the displacement field or the force vector acting on the boundary of the domain.

An elastic body under the action of external forces experiences deformation, mathematically described by means of the stress and strain tensors and the displacement vector (see [9, 1]). These quantities satisfy a system of equations that, under certain conditions, can be reduced to one in only two—rather than the original three—independent space variables. Such is the case of the system governing the equilibrium of thin elastic plates with transverse shear deformation, extensively investigated in [6, 7] in spaces of smooth functions, and in [2] in spaces of distributions. The corresponding stationary oscillations model has been studied in [16], and that of dynamic deformations in [3].

In this book, we consider the system of equations (known as the Winkler model [18]) that describes the equilibrium of a thin elastic plate with in-plane deformation and no bending, which lies on an elastic foundation and is subjected to Dirichlet, Neumann, or Robin boundary conditions. This model has many important applications in engineering problems arising in geotechnical research, road construction, biomechanics, and other practical fields. A brief presentation of some preliminary results can be found in [4, 5]. Our intention is to describe the mathematical model analytically and then use it to show how a boundary element method, based on the boundary integral equation technique, can be constructed and manipulated to compute an approximate (numerical) solution. The advantage of this type of approach over the use of finite elements or other classical computational procedures is twofold: it reduces the original two-dimensional setup to a one-dimensional problem, and provides a faster rate of convergence.

The material in the book is organized in five chapters.

In Chapter 1, we describe the model and list the main boundary value problems associated with its governing system.

Chapter 2 contains the definition of the elastic potentials in terms of a matrix of fundamental solutions, and a description of the mapping properties of the boundary integral operators generated by these potentials.

The main thrust in Chapter 3 is the use of the layer potentials in the solution of our boundary value problems by means of direct and classical indirect BIEMs. These results are fundamental in establishing the well-posed nature of the mathematical model.

In Chapter 4, we make a detailed presentation of the *Mathematica*® software in relation to its use in our numerical handling of the problems.

Finally, Chapter 5 consists of a collection of computational examples that illustrate the implementation of the collocation method for the direct and classical indirect BIEMs with various conditions prescribed on smooth and non-smooth boundaries, and different choices of splines.

The book should be a good source of information for readers who want to get at-a-glance theoretical and practical details about the analytic structure and numerical applications (in the *Mathematica*® environment) of boundary integral equation methods.

We wish to express our thanks to Elizabeth Loew, Executive Editor for Mathematics at Springer, who has guided the evolution of this project with impeccable professionalism and great efficiency.

Also, two of us (CC and DD) are grateful to our wives for their support, understanding, and remarkable staying power in the face of stiff competition from our computers during the production of this book.

Tulsa, OK, USA   Christian Constanda
Tulsa, OK, USA   Dale Doty
Tulsa, OK, USA   William Hamill
September 2015

Contents

Chapter 1
The Mathematical Model

1.1 Basic Equations

Throughout the book, Latin and Greek subscripts take the values $1,2,3$ and $1,2$, respectively, and the convention of summation over repeated indices is understood. For simplicity, we denote by I both the identity matrix on any space of square matrices and the identity operator on any space of functions. Also, we denote the transpose of a matrix X by X^{T} and the derivatives of a function $f = f(x_i)$ by

$$\frac{\partial f}{\partial x_i} = f_{,i} = \partial_i f,$$

$$\frac{\partial^2 f}{\partial x_i \partial x_j} = f_{,ij} = \partial_i \partial_j f,$$

with the obvious generalization for higher-order derivatives.

Let S be a finite domain in $\mathbb{R}^2$, with a simple and closed boundary ∂S oriented in the positive direction.

1.1 Definition. (i) A function f defined in S (on ∂S) is called *Hölder continuous with index* $\alpha \in (0,1)$ if there is $c = \text{const} > 0$ such that

$$|f(x) - f(y)| \leq c|x - y|^{\alpha} \quad \forall x, y \in S \ (\forall x, y \in \partial S).$$

The space of all such functions is denoted by $C^{0,\alpha}(S)$ ($C^{0,\alpha}(\partial S)$).

(ii) A function f defined in S (on ∂S) is called *Hölder continuously differentiable with index* $\alpha \in (0,1)$ if it is differentiable in S (on ∂S) and its derivatives belong to $C^{0,\alpha}(S)$ ($C^{0,\alpha}(\partial S)$). The space of all such functions is denoted by $C^{1,\alpha}(S)$ ($C^{1,\alpha}(\partial S)$).

1.2 Remark. If S is an infinite domain, then a function f defined on S is Hölder continuous (Hölder continuously differentiable) in S if it has this property on any finite subdomain of S. Of course, the constant c in Definition 1.1 will vary with the subdomain.

© Springer International Publishing Switzerland 2016
C. Constanda et al., *Boundary Integral Equation Methods and Numerical Solutions*,
Developments in Mathematics 35, DOI 10.1007/978-3-319-26309-0_1

We now assume that the boundary curve ∂S is of class C^2, denote S by S^+, and write

$$S^- = \mathbb{R}^2 \setminus (S^+ \cup \partial S), \quad \bar{S}^+ = S^+ \cup \partial S, \quad \bar{S}^- = S^- \cup \partial S.$$

Let $x = (x_1, x_2)$ be a generic point in a Cartesian system of coordinates with the origin $(0,0) \in S^+$, and let $v(x)$ be the outward unit normal at x on ∂S (see Fig. 1.1). For $x, y \in \mathbb{R}^2$, we write

$$|x - y| = [(x_1 - y_1)^2 + (x_2 - y_2)^2]^{1/2}.$$

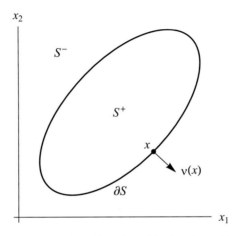

Fig. 1.1 The geometric configuration of the domains and boundary.

The equilibrium equations of three-dimensional elasticity are

$$t_{ij,j} + q_i = 0,$$

where the $t_{ij} = t_{ji}$ are the components of the stress tensor and q_i are the components of the body force vector. For a homogeneous and isotropic material, the constitutive equations have the form

$$t_{ij} = \lambda u_{k,k} \delta_{ij} + \mu (u_{i,j} + u_{j,i}),$$

where u_i are the displacement components, λ and μ are the elastic (Lamé) coefficients, and δ_{ij} are the Kronecker delta. The components of the stress vector in a direction $n = (n_1, n_2, n_3)^{\mathrm{T}}$ are

$$t_i = t_{ij} n_j,$$

and the internal energy density is

$$\mathscr{E} = \tfrac{1}{4} t_{ij} (u_{i,j} + u_{j,i}) = \tfrac{1}{2} t_{ij} u_{i,j}.$$

The model of plain strain is based on the assumption that

$$u_\alpha = u_\alpha(x_1, x_2), \quad u_3 = 0,$$
$$q_\alpha = q_\alpha(x_1, x_2), \quad q_3 = 0,$$
$$n = (n_1, n_2, 0)^{\mathrm{T}}.$$

Hence, the equilibrium equations, the constitutive relations, the stress tensor components, and the internal energy density can be expressed as

$$t_{\alpha\beta,\beta} + q_\alpha = 0, \tag{1.1}$$

$$t_{\alpha\beta} = \lambda u_{\gamma,\gamma} \delta_{\alpha\beta} + \mu(u_{\alpha,\beta} + u_{\beta,\alpha}), \tag{1.2}$$

$$t_\alpha = t_{\alpha\beta} n_\beta, \tag{1.3}$$

$$\mathscr{E} = \tfrac{1}{2} t_{\alpha\beta} u_{\alpha,\beta}. \tag{1.4}$$

Combining (1.1) and (1.2), we find that

$$(\lambda + \mu) u_{\beta,\beta\alpha} + \mu u_{\alpha,\beta\beta} + q_\alpha = 0,$$

or

$$(\lambda + \mu)\operatorname{grad}\operatorname{div} u + \mu \Delta u = q, \tag{1.5}$$

which can be written as

$$A(\partial_1, \partial_2) u = q,$$

where

$$u = (u_1, u_2)^{\mathrm{T}}, \quad q = (q_1, q_2)^{\mathrm{T}}, \quad \Delta = \partial_1^2 + \partial_2^2,$$

and $A(\partial_1, \partial_2)$ is the matrix differential operator

$$A(\partial_1, \partial_2) = \begin{pmatrix} \mu\Delta + (\lambda + \mu)\partial_1^2 & (\lambda + \mu)\partial_1\partial_2 \\ (\lambda + \mu)\partial_1\partial_2 & \mu\Delta + (\lambda + \mu)\partial_2^2 \end{pmatrix}.$$

In the homogeneous case—that is, in the absence of body forces—system (1.5) becomes

$$A(\partial_1, \partial_2) u = 0.$$

If the plate lies on an elastic foundation, the differential operator changes to

$$Z(\partial_1, \partial_2) = \begin{pmatrix} \mu\Delta + (\lambda + \mu)\partial_1^2 - k & (\lambda + \mu)\partial_1\partial_2 \\ (\lambda + \mu)\partial_1\partial_2 & \mu\Delta + (\lambda + \mu)\partial_2^2 - k \end{pmatrix}, \tag{1.6}$$

or

$$Z(\partial_1, \partial_2) = A(\partial_1, \partial_2) - kI, \tag{1.7}$$

where $k = \text{const} > 0$ is the elastic constant of the foundation material. Then the homogeneous equilibrium system can be written as

$$Z(\partial_1, \partial_2)u = 0. \tag{1.8}$$

1.3 Theorem. *If*

$$\lambda + \mu > 0, \quad \mu > 0, \tag{1.9}$$

then system (1.8) *is elliptic.*

Proof. Since, under conditions (1.9), the system $A(\partial_1, \partial_2)u = 0$ is elliptic (see Theorem 2.1 in [6]), from (1.7) it follows that so is (1.8). $\qquad\square$

1.4 Remark. In what follows, we assume that inequalities (1.9) are satisfied.

Combining (1.2) and (1.3) and setting $n_\alpha = v_\alpha$ (the components of the outward unit normal to ∂S), we arrive at

$$t_\alpha = t_{\alpha\beta} v_\beta = T_{\alpha\beta} u_\beta,$$

where the boundary stress operator $T(\partial_1, \partial_2)$ is given by

$$T(\partial_1, \partial_2) = \begin{pmatrix} (\lambda + 2\mu)\,v_1\partial_1 + \mu\,v_2\partial_2 & \mu v_2\partial_1 + \lambda v_1\partial_2 \\ \lambda v_2\partial_1 + \mu v_1\partial_2 & \mu\,v_1\partial_1 + (\lambda + 2\mu)\,v_2\partial_2 \end{pmatrix}. \tag{1.10}$$

From (1.2) and (1.4) it follows that the internal energy density in the absence of a foundation can be written as

$$\mathscr{E} = \mathscr{E}(u,u) = \tfrac{1}{2}[(\lambda + 2\mu)(u_{1,1}^2 + u_{2,2}^2) + 2\lambda u_{1,1} u_{2,2} + \mu(u_{1,2} + u_{2,1})^2].$$

For the elastic foundation problem, the internal energy density is

$$E = E(u,u) = \mathscr{E}(u,u) + \tfrac{1}{2}k u^\mathrm{T} u. \tag{1.11}$$

1.5 Theorem. $E(u,u)$ *is a positive definite quadratic form.*

This assertion follows from Theorem 2.2 in [6], which states that $\mathscr{E}(u,u)$ is positive definite, and the fact that the second term in (1.11) is also positive definite.

1.2 Boundary Value Problems

Let $\mathscr{M}_{p\times q}$ be the vector space of $p \times q$ matrices, and let $\mathscr{A}$ be the vector space of functions $u \in \mathscr{M}_{2\times 1}$ in S^- such that, in polar coordinates,

$$u(r,\theta) = O(r^{-1-\alpha}), \quad \alpha > 0, \quad \text{as } r \to \infty. \tag{1.12}$$

Let $\mathscr{P}, \mathscr{R}, \mathscr{Q}, \mathscr{S}, \mathscr{K}, \mathscr{L} \in C(\partial S)$ be given 2×1 matrix-valued functions, and let $\sigma \in C(\partial S)$ be a positive definite 2×2 matrix-valued function. We consider the following six basic boundary value problems:

(i) Interior Dirichlet problem (D$^+$): Find $u \in C^2(S^+) \cap C^1(\bar{S}^+)$ such that

$$(Zu)(x) = 0, \quad x \in S^+, \quad u(x) = \mathscr{P}(x), \quad x \in \partial S.$$

(ii) Exterior Dirichlet problem (D$^-$): Find $u \in C^2(S^-) \cap C^1(\bar{S}^-) \cap \mathscr{A}$ such that

$$(Zu)(x) = 0, \quad x \in S^-, \quad u(x) = \mathscr{R}(x), \quad x \in \partial S.$$

(iii) Interior Neumann problem (N$^+$): Find $u \in C^2(S^+) \cap C^1(\bar{S}^+)$ such that

$$(Zu)(x) = 0, \quad x \in S^+, \quad (Tu)(x) = \mathscr{Q}(x), \quad x \in \partial S.$$

(iv) Exterior Neumann problem (N$^-$): Find $u \in C^2(S^-) \cap C^1(\bar{S}^-) \cap \mathscr{A}$ such that

$$(Zu)(x) = 0, \quad x \in S^-, \quad (Tu)(x) = \mathscr{S}(x), \quad x \in \partial S.$$

(v) Interior Robin problem (R$^+$): Find $u \in C^2(S^+) \cap C^1(\bar{S}^+)$ such that

$$(Zu)(x) = 0, \quad x \in S^+, \quad (Tu + \sigma u)(x) = \mathscr{K}(x), \quad x \in \partial S.$$

(vi) Exterior Robin Problem (R$^-$): Find $u \in C^2(S^-) \cap C^1(\bar{S}^-) \cap \mathscr{A}$ such that

$$(Zu)(x) = 0, \quad x \in S^-, \quad (Tu - \sigma u)(x) = \mathscr{L}(x), \quad x \in \partial S.$$

Any function that satisfies pointwise one of the above systems of equations is termed a *regular solution*, or, simply, a *solution* of that boundary value problem.

The following assertion is analogous to Green's first identity [13].

1.6 Theorem. *If $u \in C^2(S^+) \cap C^1(\bar{S}^+)$, then*

$$\int_{S^+} u^{\mathrm{T}}(Zu)\, da = \int_{\partial S} u^{\mathrm{T}}(Tu)\, ds - 2 \int_{S^+} E(u,u)\, da.$$

Proof. From (1.7) and Theorem 2.7 in [6] it follows that

$$\int_{S^+} u^{\mathrm{T}}(Zu)\, da = \int_{S^+} u^{\mathrm{T}}(A - kI)u\, da = \int_{S^+} u^{\mathrm{T}}(Au)\, da - \int_{S^+} u^{\mathrm{T}}(ku)\, da$$

$$= \int_{\partial S} u^{\mathrm{T}}(Tu)\, ds - 2 \int_{S^+} \mathscr{E}(u,u)\, da - k \int_{S^+} u^{\mathrm{T}} u\, da$$

$$= \int_{\partial S} u^{\mathrm{T}}(Tu)\, ds - 2 \left(\int_{S^+} \mathscr{E}(u,u)\, da + \tfrac{1}{2} k \int_{S^+} u^{\mathrm{T}} u\, da \right)$$

$$= \int_{\partial S} u^{\mathrm{T}}(Tu)\, ds - 2 \int_{S^+} E(u,u)\, da. \qquad \Box$$

The above assertion has an important consequence.

1.7 Corollary (Betti formulas). (i) *If* $u \in C^2(S^+) \cap C^1(\bar{S}^+)$ *and* $Zu = 0$ *in* S^+, *then*

$$2 \int_{S^+} E(u,u)\,da = \int_{\partial S} u^{\mathrm{T}}(Tu)\,ds. \qquad (1.13)$$

(ii) *If* $u \in C^2(S^-) \cap C^1(\bar{S}^-) \cap \mathscr{A}$ *and* $Zu = 0$ *in* S^-, *then*

$$2 \int_{S^-} \mathscr{E}(u,u)\,da = -\int_{\partial S} u^{\mathrm{T}}(Tu)\,ds. \qquad (1.14)$$

Proof. (i) By Theorem 1.6,

$$\int_{S^+} u^{\mathrm{T}}(Zu)\,da = \int_{\partial S} (Tu)\,ds - 2\int_{S^+} E(u,u)\,da,$$

and the result follows from the fact that $Zu = 0$ in S^+.

(ii) Let K_R be a disc centered at the origin and of radius R sufficiently large so that $\bar{S}^+ \subset K_R$ (see Fig. 1.2).

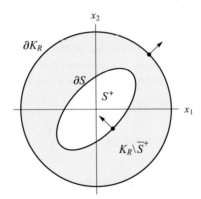

Fig. 1.2 The domain $K_R \setminus \bar{S}^+$.

By (1.7),

$$\int_{K_R \setminus \bar{S}^+} u^{\mathrm{T}}(Zu)\,da = \int_{K_R \setminus \bar{S}^+} u^{\mathrm{T}}(Au)\,da - k \int_{K_R \setminus \bar{S}^+} u^{\mathrm{T}}u\,da.$$

Since $Zu = 0$ in $K_R \setminus \bar{S}^+ \subset S^-$, it follows that

$$0 = \int_{K_R \setminus \bar{S}^+} u^{\mathrm{T}}(Au)\,da - k \int_{K_R \setminus \bar{S}^+} u^{\mathrm{T}}u\,da.$$

Then, by Corollary 2.8 in [6],

$$0 = \int_{\partial(K_R \setminus S^+)} u^{\mathrm{T}}(Tu)\,ds - 2 \int_{K_R \setminus S^+} \mathscr{E}(u,u)\,da - k \int_{K_R \setminus S^+} u^{\mathrm{T}} u\,da. \qquad (1.15)$$

Since here the outward normal to ∂S is directed into S^+ (see Fig. 1.2), we have

$$\int_{\partial(K_R \setminus \bar{S}^+)} u^{\mathrm{T}}(Tu)\,ds = \int_{\partial K_R} u^{\mathrm{T}}(Tu)\,ds - \int_{\partial S} u^{\mathrm{T}}(Tu)\,ds,$$

which, when substituted in (1.15), yields

$$0 = \int_{\partial K_R} u^{\mathrm{T}}(Tu)\,ds - \int_{\partial S} u^{\mathrm{T}}(Tu)\,ds$$

$$- 2 \int_{K_R \setminus S^+} \mathscr{E}(u,u)\,da - k \int_{K_R \setminus S^+} u^{\mathrm{T}} u\,da. \qquad (1.16)$$

By (1.12),

$$u = O(r^{-1-\alpha}), \quad \alpha > 0,$$

so

$$u^{\mathrm{T}} u = u_1^2 + u_2^2 = O(R^{-2-2\alpha}) \quad \text{in } K_R \setminus \bar{S}^+;$$

therefore,

$$\int_{K_R \setminus S^+} u^{\mathrm{T}} u\,da = O(R^{-2-2\alpha})O(R^2) = O(R^{-2\alpha}) \to 0 \quad \text{as } R \to \infty. \qquad (1.17)$$

Since

$$\partial_1 u, \ \partial_2 u = O(r^{-2-\alpha}),$$

it follows that

$$\int_{\partial K_R} u^{\mathrm{T}}(Tu)\,ds = O(R^{-1-\alpha})O(R^{-2-\alpha})O(R)$$

$$= O(R^{-2-2\alpha}) \to 0 \quad \text{as } R \to \infty. \qquad (1.18)$$

Also, $K_R \setminus S^+$ expands into the whole of $\bar{S}^-$ as $R \to \infty$; therefore, in the limit, formulas (1.16)–(1.18) generate (1.14). $\qquad \square$

The analog of Green's second identity [13] is called the *reciprocity relation*.

1.8 Theorem. *For any $u, v \in C^2(S^+) \cap C^1(\bar{S}^+)$,*

$$\int_{S^+} [u^{\mathrm{T}}(Zv) - v^{\mathrm{T}}(Zu)]\,da = \int_{\partial S} [u^{\mathrm{T}}(Tv) - v^{\mathrm{T}}(Tu)]\,ds.$$

Proof. By Theorem 2.9 in [6],

$$\int_{S^+} [u^\mathrm{T}(Zv) - v^\mathrm{T}(Zu)]\, da = \int_{S^+} [u^\mathrm{T}(A - kI)v - v^\mathrm{T}(A - kI)u]\, da$$

$$= \int_{S^+} [u^\mathrm{T}(Av) - v^\mathrm{T}(Au)]\, da - k \int_{S^+} (u^\mathrm{T}v - v^\mathrm{T}u)\, da$$

$$= \int_{S^+} [u^\mathrm{T}(Av) - v^\mathrm{T}(Au)]\, da = \int_{\partial S} [u^\mathrm{T}(Tv) - v^\mathrm{T}(Tu)]\, ds. \quad \square$$

1.9 Theorem (Uniqueness). *Each of (D^+), (D^-), (N^+), (N^-), (R^+), and (R^-) has at most one solution.*

Proof. Let $u = v - w$, where $v, w \in C^2(S^+) \cap C^1(\bar{S}^+)$ are any two solutions of one of (D^+) or (N^+) with the same nonhomogeneous term. Then u is a solution of the corresponding homogeneous problem, so, by (1.13),

$$2 \int_{S^+} E(u,u)\, da = \int_{\partial S} u^\mathrm{T}(Tu)\, ds = 0.$$

Since $E(u,u)$ is a positive definite quadratic form, it follows that $E(u,u) = 0$, which implies that $u = 0$, or $v = w$; hence, each of (D^+) and (N^+) has at most one solution.

The proof for (D^-) and (N^-) is similar, with (1.14) used instead of (1.13).

We now turn to the Robin problems. As above, the difference $u = v - w$ of any two solutions $v, w \in C^2(S^+) \cap C^1(\bar{S}^+)$ of (R^+) for a given $\mathscr{K}$ and a given positive definite 2×2 matrix function $\sigma \in C(\partial S)$, is a solution of the homogeneous problem (R^+), so, again, from (1.13) it follows that

$$2 \int_{S^+} E(u,u)\, da = \int_{\partial S} u^\mathrm{T}(Tu)\, ds = - \int_{\partial S} u^\mathrm{T}(\sigma u)\, ds,$$

or, what is the same,

$$2 \int_{S^+} E(u,u)\, da + \int_{\partial S} u^\mathrm{T}(\sigma u)\, ds = 0.$$

We have

$$u^\mathrm{T}\sigma u = \sigma_{\alpha\beta} u_\alpha u_\beta;$$

consequently, since $E(u,u)$ is a positive definite quadratic form and σ is a positive definite matrix, we conclude that u is a rigid displacement in S^+ that vanishes on ∂S, so $u = 0$, or $v = w$, which means that (R^+) has at most one solution.

The proof for (R^-) is similar to that for (R^+), with the Betti formula (1.14) instead of (1.13). $\quad \square$

Chapter 2
The Layer Potentials

2.1 Fundamental Solutions

We need to construct a matrix of fundamental solutions for the operator $Z(\partial_1, \partial_2)$. This is a two-point 2×2 matrix function satisfying

$$Z(\partial_x)D(x,y) = -\delta(|x-y|)I, \tag{2.1}$$

where ∂_x indicates that differentiation is applied with respect to the point x and δ is the Dirac delta distribution. If the scalar two-point function $t(x,y)$ is a solution of the equation

$$(\det Z)(\partial_x)t(x,y) = -\delta(|x-y|),$$

then

$$Z(\partial_x)[(\operatorname{adj} Z)(\partial_x)t(x,y)I] = (\det Z)(\partial_x)t(x,y)I = -\delta(|x-y|)I,$$

so

$$D(x,y) = (\operatorname{adj} Z)(\partial_x)t(x,y)I \tag{2.2}$$

is a matrix of fundamental solutions for Z.

We have

$$
\begin{aligned}
(\det Z)(\partial_x) &= \begin{vmatrix} \mu\Delta + (\lambda+\mu)\partial_1^2 - k & (\lambda+\mu)\partial_1\partial_2 \\ (\lambda+\mu)\partial_1\partial_2 & \mu\Delta + (\lambda+\mu)\partial_2^2 - k \end{vmatrix} \\
&= \mu(\lambda+2\mu)\Delta^2 - 2k\mu\Delta - (\lambda+\mu)k\Delta + k^2 \\
&= \mu(\lambda+2\mu)\left(\Delta^2 - \frac{k(\lambda+3\mu)}{\mu(\lambda+2\mu)}\Delta + \frac{k^2}{\mu(\lambda+2\mu)}\right) \\
&= \mu(\lambda+2\mu)(\Delta - C_1^2)(\Delta - C_2^2),
\end{aligned}
\tag{2.3}
$$

where

$$C_1^2 = \frac{k}{\mu}, \quad C_2^2 = \frac{k}{\lambda+2\mu}.$$

© Springer International Publishing Switzerland 2016
C. Constanda et al., *Boundary Integral Equation Methods and Numerical Solutions*,
Developments in Mathematics 35, DOI 10.1007/978-3-319-26309-0_2

A fundamental solution for the operator $\Delta - h^2 I$, $h = $ const, is [17]

$$-(1/(2\pi))K_0(h|x-y|),$$

where K_0 is the modified Bessel function of the second kind and order zero (see Fig. 2.1); that is,

$$(\Delta - h^2)K_0(h|x-y|) = -2\pi\delta(|x-y|).$$

$K_0(r)$

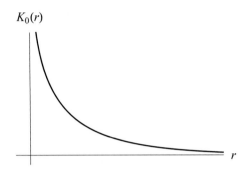

Fig. 2.1 The modified Bessel function K_0.

This suggests that we seek $t(x,y)$ of the form

$$t(x,y) = pK_0(C_1|x-y|) + qK_0(C_2|x-y|), \quad p,q = \text{const},$$

which then leads to

$$
\begin{aligned}
(\Delta - C_2^2)&t(x,y)\\
&= p(\Delta - C_2^2)K_0(C_1|x-y|) + q(\Delta - C_2^2)K_0(C_2|x-y|)\\
&= p(\Delta - C_2^2)K_0(C_1|x-y|) - 2\pi q\delta(|x-y|)\\
&= p[(\Delta - C_1^2) + (C_1^2 - C_2^2)]K_0(C_1|x-y|) - 2\pi q\delta(|x-y|)\\
&= p(\Delta - C_1^2)K_0(C_1|x-y|) + p(C_1^2 - C_2^2)K_0(C_1|x-y|) - 2\pi q\delta(|x-y|)\\
&= -2\pi p\delta(|x-y|) + p(C_1^2 - C_2^2)K_0(C_1|x-y|) - 2\pi q\delta(|x-y|).
\end{aligned}
$$

To eliminate δ, we take $p = -q$; hence,

$$(\Delta - C_2^2)t(x,y) = p(C_1^2 - C_2^2)K_0(C_1|x-y|),$$

so

$$
\begin{aligned}
(\Delta - C_1^2)(\Delta - C_2^2)t(x,y) &= (\Delta - C_1^2)p(C_1^2 - C_2^2)K_0(C_1|x-y|)\\
&= p(C_1^2 - C_2^2)(\Delta - C_1^2)K_0(C_1|x-y|)\\
&= -2\pi p(C_1^2 - C_2^2)\delta(|x-y|). \quad (2.4)
\end{aligned}
$$

Using (2.3) and (2.4), we find that

$$
\begin{aligned}
(\det Z)(\partial_x)t(x,y) &= \mu(\lambda+2\mu)(\Delta-C_1^2)(\Delta-C_2^2)t(x,y) \\
&= -2\pi\mu(\lambda+2\mu)A(C_1^2-C_2^2)\delta(|x-y|) \\
&= -2\pi p\mu(\lambda+2\mu)\left(\frac{k}{\mu}-\frac{k}{\lambda+2\mu}\right)\delta(|x-y|) \\
&= -2\pi pk(\lambda+\mu)\delta(|x-y|) \\
&= -\delta(|x-y|),
\end{aligned}
$$

where

$$
p = \frac{1}{2\pi k(\lambda+\mu)},
$$
$$
q = -\frac{1}{2\pi k(\lambda+\mu)};
$$

consequently,

$$
t(x,y) = \frac{1}{2\pi k(\lambda+\mu)}\left[K_0(C_1|x-y|)-K_0(C_2|x-y|)\right]. \tag{2.5}
$$

In this case,

$$
\begin{aligned}
D(x,y) &= (\operatorname{adj}Z)(\partial_x)t(x,y)I \\
&= \begin{pmatrix} \mu\Delta+(\lambda+\mu)\partial_2^2-k & -(\lambda+\mu)\partial_1\partial_2 \\ -(\lambda+\mu)\partial_1\partial_2 & \mu\Delta+(\lambda+\mu)\partial_1^2-k \end{pmatrix}\begin{pmatrix} t(x,y) & 0 \\ 0 & t(x,y) \end{pmatrix}, \tag{2.6}
\end{aligned}
$$

or, component-wise,

$$
\begin{aligned}
D_{11}(x,y) &= \mu\Delta t(x,y)+(\lambda+\mu)\partial_2^2 t(x,y)-kt(x,y), \\
D_{12}(x,y) &= D_{21}(x,y) = -(\lambda+\mu)\partial_1\partial_2 t(x,y), \tag{2.7} \\
D_{22}(x,y) &= \mu\Delta t(x,y)+(\lambda+\mu)\partial_1^2 t(x,y)-kt(x,y),
\end{aligned}
$$

with all differentiation performed with respect to the point x.

We notice the symmetry

$$
D(x,y) = D(y,x) = D^{\mathrm{T}}(x,y). \tag{2.8}
$$

Another necessary tool for handling our boundary value problems is the matrix of singular solutions

$$
P(x,y) = [T(\partial_y)D(y,x)]^{\mathrm{T}}, \tag{2.9}
$$

where

$$P_{11}(x,y) = \mu(\lambda + 2\mu)\partial_{v(y)}\Delta(y)t(x,y) + 2\mu(\lambda + \mu)\partial_{s(y)}t_{,12}(x,y)$$
$$- k[v_1(y)(\lambda + 2\mu)t_{,1}(x,y) + v_2(y)t_{,2}(x,y)],$$

$$P_{12}(x,y) = \mu\partial_{s(y)}\Delta(y)t(x,y) - 2\mu(\lambda + \mu)\partial_{s(y)}t_{,11}(x,y)$$
$$- k[v_1(y)t_{,2}(x,y) + v_2(y)t_{,1}(x,y)],$$

$$P_{21}(x,y) = -\mu\partial_{s(y)}\Delta(y)t(x,y) + 2\mu(\lambda + \mu)\partial_{s(y)}t_{,22}(x,y) \qquad (2.10)$$
$$- k[v_1(y)t_{,2}(x,y) + v_2(y)t_{,1}(x,y)],$$

$$P_{22}(x,y) = \mu(\lambda + 2\mu)\partial_{v(y)}\Delta(y)t(x,y) - 2\mu(\lambda + \mu)\partial_{s(y)}t_{,12}(x,y)$$
$$- k[v_1(y)(\lambda + 2\mu)t_{,1}(x,y) - v_2(y)t_{,2}(x,y)];$$

here, we have denoted the normal and tangential derivatives at y on ∂S by

$$\partial_{v(y)} = \partial/\partial v(y) = v_1(y)\partial_1 + v_2(y)\partial_2,$$
$$\partial_{s(y)} = \partial/\partial s(y) = -v_2(y)\partial_1 + v_1(y)\partial_2,$$

with differentiation in terms of the point y.

2.1 Theorem. *For $x \neq y$, the columns $D^{(\alpha)}$ of D and the columns $P^{(\alpha)}$ of P are solutions of (1.8).*

Proof. The first part of the assertion follows from the definition of D in (2.1). In fact, we can write

$$Z(\partial_x)D(x,y) = 0, \quad x \neq y. \qquad (2.11)$$

Furthermore, in view of the symmetry (2.8), we also have

$$Z(\partial_y)D(x,y) = 0, \quad Z(\partial_x)D(y,x) = 0, \quad x \neq y.$$

Finally, by (2.11) and (2.9), we see that for $x \neq y$,

$$[Z(\partial_x)P^{(\alpha)}(x,y)]_\beta = Z_{\beta\gamma}(\partial_x)P_\gamma^{(\alpha)}(x,y)$$
$$= Z_{\beta\gamma}(\partial_x)P_{\gamma\alpha}(x,y)$$
$$= Z_{\beta\gamma}(\partial_x)[T(\partial_y)D(y,x)]_{\alpha\gamma}$$
$$= Z_{\beta\gamma}(\partial_x)[T_{\alpha\sigma}(\partial_y)D_{\sigma\gamma}(y,x)]$$
$$= Z_{\beta\gamma}(\partial_x)[T_{\alpha\sigma}(\partial_y)D_{\gamma\sigma}(x,y)]$$
$$= T_{\alpha\sigma}(\partial_y)[Z_{\beta\gamma}(\partial_x)D_{\gamma\sigma}(x,y)]$$
$$= T_{\alpha\sigma}(\partial_y)[Z(\partial_x)D(x,y)]_{\beta\sigma}$$
$$= 0. \qquad \square$$

It is important to know the asymptotic behavior of $D(x,y)$ and $P(x,y)$ for small and large values of $|x - y|$.

First, for x close to y we have [6]

$$K_0(\xi) = -\left(1 + \tfrac{1}{4}\xi^2 + \tfrac{1}{64}\xi^4 + \cdots\right)\ln\xi,$$

so

$$K_0(C_1|x-y|) - K_0(C_2|x-y|)$$
$$= \tfrac{1}{4}(C_2^2 - C_1^2)r^2\ln|x-y| + \tfrac{1}{64}(C_2^4 - C_1^4)|x-y|^4\ln|x-y| + \cdots$$
$$+ \tfrac{1}{4}(C_2^2\ln C_2 - \gamma_1^2\ln C_1)|x-y|^2 + \tfrac{1}{64}(C_2^4\ln C_2 - C_1^4\ln C_1)|x-y|^4 + \cdots,$$

and, by (2.5),

$$t(x,y) = A|x-y|^2\ln(|x-y|) + B|x-y|^2 + O(|x-y|^4\ln|x-y|), \qquad (2.12)$$

where A and B are combinations of λ and μ. Direct computation now shows that, in terms of the point x,

$$|x-y|_{,\alpha} = \frac{x_\alpha - y_\alpha}{|x-y|},$$

$$|x-y|_{,\alpha\beta} = \frac{1}{|x-y|}\delta_{\alpha\beta} - \frac{(x_\alpha - y_\alpha)(x_\beta - y_\beta)}{|x-y|^3},$$

$$(\ln|x-y|)_{,\alpha} = \frac{x_\alpha - y_\alpha}{|x-y|^2},$$

$$(\ln|x-y|)_{,\alpha\beta} = \frac{1}{|x-y|^2}\delta_{\alpha\beta} - 2\frac{(x_\alpha - y_\alpha)(x_\beta - y_\beta)}{|x-y|^4},$$

$$(|x-y|^2\ln|x-y|)_{,\alpha\beta} = (2\ln|x-y| + 1)\delta_{\alpha\beta} + 2\frac{(x_\alpha - y_\alpha)(x_\beta - y_\beta)}{|x-y|^2};$$

consequently, from (2.7) and (2.12) it follows that

$$D(x,y) = \begin{pmatrix} A\ln|x-y| - B\dfrac{(x_1-y_1)^2}{|x-y|^2} & C\dfrac{(x_1-y_1)(x_2-y_2)}{|x-y|^2} \\[2ex] C\dfrac{(x_1-y_1)(x_2-y_2)}{|x-y|^2} & A\ln|x-y| - B\dfrac{(x_2-y_2)^2}{|x-y|^2} \end{pmatrix}$$

$$+ O(|x-y|^2\ln|x-y|), \quad (2.13)$$

where C is another combination of λ and μ. This indicates that for x close to y,

$$D(x,y) = O(\ln|x-y|). \qquad (2.14)$$

A similar computation, but with differentiation in terms of the point y, leads to the expansions

$$P_{11}(x,y) = v_1(y)\left(a_{11}\frac{y_1 - x_1}{|y-x|^2} + b_{11}\frac{(y_1 - x_1)(y_2 - x_2)^2}{|y-x|^4}\right.$$
$$\left. + c_{11}\frac{(y_1 - x_1)^3 - (y_1 - x_1)(y_2 - x_2)^3}{|y-x|^3} + \cdots\right)$$
$$+ v_2(y)\left(d_{11}\frac{y_2 - x_2}{|y-x|^2} + e_{11}\frac{(y_1 - x_1)^2(y_2 - x_2)}{|y-x|^4}\right.$$
$$\left. + f_{11}\frac{(y_1 - x_1)^2(y_2 - x_2) - (y_2 - x_2)^3}{|y-x|^4} + \cdots\right),$$

$$P_{12}(x,y) = v_1(y)\left(a_{12}\frac{y_2 - x_2}{|y-x|^2} + b_{12}\frac{(y_1 - x_1)^2(y_2 - x_2)}{|y-x|^4}\right.$$
$$\left. + c_{12}\frac{(y_2 - x_2)(y_1 - x_1)^2 - (y_2 - x_2)^3}{|y-x|^4} + \cdots\right)$$
$$+ v_2(y)\left(d_{12}\frac{y_1 - x_1}{|y-x|^2} + e_{12}\frac{(y_1 - x_1)(y_2 - x_2)^2}{|y-x|^4}\right.$$
$$\left. + f_{12}\frac{(y_1 - x_1)^3 - (y_1 - x_1)(y_2 - x_2)^2}{|y-x|^4} + \cdots\right),$$

$$P_{21}(x,y) = v_1(y)\left(a_{21}\frac{y_2 - x_2}{|y-x|^2} + b_{21}\frac{(y_1 - x_1)^2(y_2 - x_2)}{|y-x|^4}\right.$$
$$\left. + c_{21}\frac{(y_2 - x_2)(y_1 - x_1)^2 - (y_2 - x_2)^3}{|y-x|^4} + \cdots\right)$$
$$+ v_2(y)\left(d_{21}\frac{y_1 - x_1}{|y-x|^2} + e_{21}\frac{(y_1 - x_1)(y_2 - x_2)^2}{|y-x|^4}\right.$$
$$\left. + f_{21}\frac{(y_1 - x_1)^3 - (y_1 - x_1)(y_2 - x_2)^2}{|y-x|^4} + \cdots\right),$$

$$P_{22}(x,y) = v_1(y)\left(a_{22}\frac{y_1 - x_1}{|y-x|^2} + b_{22}\frac{(y_1 - x_1)(y_2 - x_2)^2}{|y-x|^4}\right.$$
$$\left. + c_{22}\frac{(y_1 - x_1)^3 - (y_1 - x_1)(y_2 - x_2)^3}{|y-x|^3} + \cdots\right)$$
$$+ v_2(y)\left(d_{22}\frac{y_2 - x_2}{|y-x|^2} + e_{22}\frac{(y_1 - x_1)^2(y_2 - x_2)}{|y-x|^4}\right.$$
$$\left. + f_{22}\frac{(y_1 - x_1)^2(y_2 - x_2) - (y_2 - x_2)^3}{|y-x|^4} + \cdots\right),$$

$$\tag{2.15}$$

where a, b, c, and d with various subscripts are combinations of λ and μ. These expansions show that for x close to y,

$$P(x,y) = O(|x-y|^{-1}). \tag{2.16}$$

On the other hand, for y on ∂S and $|x| \to \infty$, we have [6]

$$K_0(h|x-y|) = O(|x|^{-1/2}e^{-|x|}) \quad \forall h = \text{const},$$

which, by (2.12), (2.7), and (2.10), yields the asymptotic relations

$$\begin{aligned} t(x,y) &= O(|x|^{-1/2}e^{-|x|}), \\ D(x,y) &= O(|x|^{-5/2}e^{-|x|}), \\ P(x,y) &= O(|x|^{-7/2}e^{-|x|}). \end{aligned} \tag{2.17}$$

The next assertion is the analog of Green's representation formula.

2.2 Theorem (Somigliana formulas). (i) *If $u \in C^2(S^+) \cap C^1(\bar{S}^+)$ and $(Zu)(x) = 0$, $x \in S^+$, then*

$$\int_{\partial S} [D(x,y)(Tu)(y) - P(x,y)u(y)]ds(y) = \begin{cases} u(x), & x \in S^+, \\ \frac{1}{2}u(x), & x \in \partial S, \\ 0, & x \in S^-. \end{cases} \tag{2.18}$$

(ii) *If $u \in C^2(S^-) \cap C^1(\bar{S}^-) \cap \mathscr{A}$ and $(Zu)(x) = 0$, $x \in S^-$, then*

$$-\int_{\partial S} [D(x,y)(Tu)(y) - P(x,y)u(y)]ds(y) = \begin{cases} 0, & x \in S^+, \\ \frac{1}{2}u(x), & x \in \partial S, \\ u(x), & x \in S^-. \end{cases} \tag{2.19}$$

Proof. (i) Consider the region $\bar{S}^+ \setminus \sigma_{x,\varepsilon}$, where $\sigma_{x,\varepsilon}$ is a disc centered at x and of radius ε, and such that $\bar{\sigma}_{x,\varepsilon} \subset S^+$ (see Fig. 2.2).

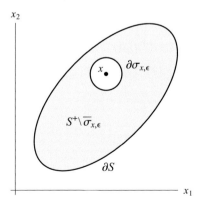

Fig. 2.2 The domain $S^+ \setminus \bar{\sigma}_{x,\varepsilon}$.

Using Theorem 1.8 in $\bar{S}^+ \setminus \sigma_{x,\varepsilon}$ with the position of u and v reversed and $v = v(y)$ replaced by the columns $D^{(\alpha)}(y,x)$ of $D(y,x)$, we find that

$$\int_{S^+\setminus\sigma_{x,\varepsilon}} [D^{(\alpha)^\mathsf{T}}(y,x)(Zu)(y) - u^\mathsf{T}(y)Z(\partial_y)D^{(\alpha)}(y,x)]\,da(y)$$

$$= \int_{\partial S} [D^{(\alpha)^\mathsf{T}}(y,x)(Tu)(y) - u^\mathsf{T}(y)T(\partial_y)D^{(\alpha)}(y,x)]\,ds(y)$$

$$+ \int_{\partial\sigma_{x,\varepsilon}} [D^{(\alpha)^\mathsf{T}}(y,x)(Tu)(y) - u^\mathsf{T}(y)T(\partial y)D^{(\alpha)}(y,x)]\,ds(y).$$

By Theorem 2.1,

$$\int_{S^+\setminus\sigma_{x,\varepsilon}} [D^{(\alpha)^\mathsf{T}}(y,x)(Zu)(y) - u^\mathsf{T}(y)Z(\partial_y)D^{(\alpha)}(y,x)]\,da(y) = 0,$$

and the above equation becomes

$$0 = \int_{\partial S} [D^{(\alpha)^\mathsf{T}}(y,x)(Tu)(y) - u^\mathsf{T}(y)T(\partial_y)D^{(\alpha)}(y,x)]\,ds(y)$$

$$+ \int_{\partial\sigma_{x,\varepsilon}} [D^{(\alpha)^\mathsf{T}}(y,x)(Tu)(y) - u^\mathsf{T}(y)T(\partial_y)D^{(\alpha)}(y,x)]\,ds(y). \quad (2.20)$$

Taking (2.9) and (2.8) into account, we can write the integrands in (2.20) as

$$D^{(\alpha)^\mathsf{T}}(y,x)(Tu)(y) - u^\mathsf{T}(y)T(\partial_y)D^{(\alpha)}(y,x)$$

$$= D_{\beta\alpha}(y,x)(Tu)_\beta(y) - u_\beta(y)(T(\partial_y)D^{(\alpha)}(y,x))_\beta$$

$$= D^\mathsf{T}_{\alpha\beta}(y,x)(Tu)_\beta(y) - (T_{\beta\gamma}(\partial_y)D_{\gamma\alpha}(y,x))u_\beta(y)$$

$$= D_{\alpha\beta}(x,y)(Tu)_\beta(y) - P_{\alpha\beta}(x,y)u_\beta(y)$$

$$= (D(x,y)(Tu)(y))_\beta - (P(x,y)u(y))_\beta,$$

so (2.20) yields

$$\int_{\partial S} [D(x,y)(Tu)(y) - P(x,y)u(y)]\,ds(y)$$

$$+ \int_{\partial\sigma_{x,\varepsilon}} [D(x,y)(Tu)(y) - P(x,y)u(y)]\,ds(y) = 0. \quad (2.21)$$

Since $u \in C^2(S^+) \cap C^1(\bar{S}^+)$, it follows that all the partial derivatives of u are bounded on ∂S, which means that Tu is bounded; therefore, by (2.14),

$$\int_{\partial \sigma_{x,\varepsilon}} D(x,y)(Tu)(y)\,ds(y) = O(\varepsilon \ln \varepsilon) \to 0 \quad \text{as } \varepsilon \to 0. \qquad (2.22)$$

Next,

$$\int_{\partial \sigma_{x,\varepsilon}} P(x,y)\,u(y)\,ds(y)$$

$$= \int_{\partial \sigma_{x,\varepsilon}} P(x,y)[u(y) - u(x)]\,ds(y) + u(x) \int_{\partial \sigma_{x,\varepsilon}} P(x,y)\,ds(y). \qquad (2.23)$$

Going over to polar coordinates with the pole at x, we represent the point $y \in \partial \sigma_{x,\varepsilon}$ as $y = (\varepsilon, \theta)$ and, taking into account the fact that the outward normal on $\partial \sigma_{x,\varepsilon}$ is directed into $\sigma_{x,\varepsilon}$, find that (2.16) has the more explicit form

$$P(x,y) = \frac{1}{2\pi\varepsilon} I + \frac{1}{\varepsilon} \mathscr{G}(\theta),$$

where

$$\int_0^{2\pi} \mathscr{G}(\theta)\,d\theta = 0.$$

Then, for $y \in \partial \sigma_{x,\varepsilon}$,

$$P(x,y) = O(\varepsilon^{-1}), \quad u(x) - u(y) = O(\varepsilon), \quad ds(y) = \varepsilon\,d\theta,$$

and

$$\int_{\partial \sigma_{x,\varepsilon}} P(x,y)[u(y) - u(x)]\,ds(y) = O(\varepsilon^{-1})O(\varepsilon)\varepsilon \to 0 \quad \text{as } \varepsilon \to 0.$$

Also,

$$u(x) \int_{\partial \sigma_{x,\varepsilon}} P(x,y)\,ds(y) = u(x) \int_{\partial \sigma_{x,\varepsilon}} [(2\pi)^{-1}\varepsilon^{-1}I + \varepsilon^{-1}\mathscr{G}(\theta)]\,\varepsilon\,d\theta$$

$$= u(x)\left(I + \int_0^{2\pi} \mathscr{G}(\theta)\,d\theta\right) = u(x).$$

Substituting this in (2.23), we conclude that

$$\int_{\partial \sigma_{x,\varepsilon}} P(x,y)u(y)\,ds(y) \to u(x) \quad \text{as } \varepsilon \to 0. \qquad (2.24)$$

The first part of (2.18) is now obtained by replacing (2.22) and (2.24) in (2.21) and letting $\varepsilon \to 0$.

If $x \in \partial S$, only one half of the disc $\sigma_{x,\varepsilon}$ is needed to isolate the point x. If $x \in S^-$, $D(x,y)$ is analytic at any point $y \in S^+$, and the argument involving the disc $\sigma_{x,\varepsilon}$ is not needed. This generates the second and third parts of (2.18).

(ii) Let K_R be a disc with the center at x and radius R sufficiently large so that $\bar{S}^+ \subset K_R$, and let $S^* = K_R \setminus \bar{S}^+$ (see Fig. 2.3). Using the first part of formula (2.18) in $\bar{S}^* = S^* \cup \partial S \cup \partial K_R$ and noting that in this case the outward normal on ∂S points into S^+, we have

$$
-\int\limits_{\partial S} [D(x,y)(Tu)(y) - P(x,y)u(y)]\,ds(y)
$$
$$
+ \int\limits_{\partial K_R} [D(x,y)(Tu)(y) - P(x,y)u(y)]\,ds(y) = u(x).
$$

In view of the asymptotic relations (2.17), the second integrand vanishes as $R \to \infty$, and we obtain the third part of (2.19). The argument for $x \in \partial S$ and $x \in S^+$ is similar to that in (i). □

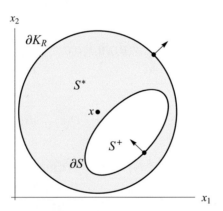

Fig. 2.3 The domain S^*.

2.2 The Layer Potentials

2.3 Definition. The functions

$$
(V\varphi)(x) = \int\limits_{\partial S} D(x,y)\varphi(y)\,ds(y),
$$
$$
(W\psi)(x) = \int\limits_{\partial S} P(x,y)\psi(y)\,ds(y),
$$

where φ and ψ are 2-component vector functions defined on ∂S, are called the *single-layer potential* and *double-layer potential* of density φ and ψ, respectively.

2.4 Theorem. *If $\varphi, \psi \in C(\partial S)$, then $V\varphi, W\psi \in \mathscr{A}$.*

Proof. The asymptotic behavior (1.12) follows immediately from (2.17). $\qquad\square$

2.5 Theorem. (i) *If $\varphi, \psi \in C(\partial S)$, then $V\varphi$ and $W\psi$ are analytic in $S^+ \cup S^-$ and*

$$Z(V\varphi) = Z(W\psi) = 0 \quad \text{in } S^+ \cup S^-.$$

(ii) *If $\varphi \in C^{0,\alpha}(\partial S)$, then $V\varphi \in C^{0,\alpha}(\mathbb{R}^2)$.*
(iii) *If $\psi \in C^{0,\alpha}(\partial S)$, then $W^{\pm}\psi \in C^{0,\alpha}(S^{\pm})$, where*

$$W^{\pm}\psi = (W\psi)|_{S^{\pm}}.$$

(iv) *If $\varphi \in C^{0,\alpha}(\partial S)$, then the direct values $V_0\varphi$ and $W_0\psi$ of $V\varphi$ and $W\psi$ on ∂S exist, with the integral in $W_0\psi$ understood as Cauchy principal value.*
(v) *The operators $\mathscr{V}^+$ and $\mathscr{V}^-$ defined by*

$$\begin{aligned} \mathscr{V}^+\varphi &= (V\varphi)|_{\bar{S}^+}, \\ \mathscr{V}^-\varphi &= (V\varphi)|_{\bar{S}^-} \end{aligned} \tag{2.25}$$

map $C^{0,\alpha}(\partial S)$ to $C^{1,\alpha}(\bar{S}^{\pm})$, $\alpha \in (0,1)$, respectively, and

$$\begin{aligned} T(\mathscr{V}^+\varphi) &= \left(W_0^* + \tfrac{1}{2}I\right)\varphi, \\ T(\mathscr{V}^-\varphi) &= \left(W_0^* - \tfrac{1}{2}I\right)\varphi, \end{aligned} \tag{2.26}$$

where W_0^*, the adjoint of the operator W_0, is defined by

$$(W_0^*\varphi)(x) = \int_{\partial S} (T(\partial_x)D(x,y))\varphi(y)\,ds(y), \quad x \in \partial S,$$

with the integral understood as Cauchy principal value.
(vi) *The operators $\mathscr{W}^+$ and $\mathscr{W}^-$ defined by*

$$\mathscr{W}^+\psi = \begin{cases} (W\psi)|_{S^+} & \text{in } S^+, \\ \left(W_0 - \tfrac{1}{2}I\right)\psi & \text{on } \partial S, \end{cases}$$

$$\mathscr{W}^-\psi = \begin{cases} (W\psi)|_{S^-} & \text{in } S^-, \\ \left(W_0 + \tfrac{1}{2}I\right)\psi & \text{on } \partial S \end{cases}$$

map $C^{0,\alpha}(\partial S)$ to $C^{0,\alpha}(\bar{S}^{\pm})$ and $C^{1,\alpha}(\partial S)$ to $C^{1,\alpha}(\bar{S}^{\pm})$, $\alpha \in (0,1)$, and

$$T(\mathscr{W}^+\psi) = T(\mathscr{W}^-\psi) \quad \text{on } C^{1,\alpha}(\partial S).$$

(vii) *The operator W_0 maps $C^{0,\alpha}(\partial S)$ to $C^{1,\alpha}(\partial S)$, $\alpha \in (0,1)$.*

Proof. Statement (i) is verified by means of Theorem 2.1.

Comparing the asymptotic expansions (2.13) and (2.15) with the corresponding ones in the absence of an elastic foundation [6], we see that they contain the same terms but with different coefficients. Since properties (ii)–(vii) depend exclusively on the order of the singularities in $D(x,y)$ and $P(x,y)$ for x close to y, their proof is a mirror image of that of their equivalents in [6]. ◻

2.6 Remarks. (i) In Theorem 2.4, the derivatives on ∂S of the functions defined on $\bar{S}^+$ and $\bar{S}^-$ are one-sided.

(ii) Clearly,

$$Z(\mathscr{V}^+\varphi)(x) = 0, \quad x \in S^+,$$
$$Z(\mathscr{V}^-\varphi)(x) = 0, \quad x \in S^-. \tag{2.27}$$

(iii) If

$$\mathscr{V}^+\varphi = \mathscr{V}^-\varphi = 0,$$

then

$$\varphi = 0. \tag{2.28}$$

This is easily seen from the fact that

$$T(\mathscr{V}^+)\varphi = T(\mathscr{V}^-)\varphi = 0,$$

and (2.28) follows from (2.26).

(iv) The Somigliana relations can be written in a more compact form in terms of the operators $\mathscr{V}^\pm$ and $\mathscr{W}^\pm$. Thus, if u is a solution of $Zu = 0$ in S^+, then

$$\mathscr{V}^+(Tu) - \mathscr{W}^+(u|_{\partial S}) = u, \quad \mathscr{V}^-(Tu) - \mathscr{W}^-(u|_{\partial S}) = 0, \tag{2.29}$$

and if u is a solution of $Zu = 0$ in S^-, then

$$-\mathscr{V}^-(Tu) + \mathscr{W}^-(u|_{\partial S}) = u, \quad -\mathscr{V}^+(Tu) + \mathscr{W}^+(u|_{\partial S}) = 0. \tag{2.30}$$

(v) From Theorem 2.5(vi) it follows that we can define a boundary operator

$$N_0 : C^{1,\alpha}(\partial S) \to C^{0,\alpha}(\partial S)$$

by setting

$$N_0\psi = T(\mathscr{W}^+\psi) = T(\mathscr{W}^-\psi), \quad \psi \in C^{1,\alpha}(\partial S). \tag{2.31}$$

(vi) The operators defined in Theorem 2.5(v),(vi) satisfy

$$(\mathscr{V}^\pm\varphi)|_{\partial S} = \mathscr{V}_0^\pm\varphi = V_0\varphi,$$
$$(\mathscr{W}^\pm\psi)|_{\partial S} = \mathscr{W}_0^\pm\psi = (W_0 \mp \tfrac{1}{2}I)\psi. \tag{2.32}$$

(vii) We note that V_0, W_0, and W_0^* are defined on $C^{0,\alpha}(\partial S)$, whereas N_0 is defined on $C^{1,\alpha}(\partial S)$, $\alpha \in (0,1)$.

2.3 Properties of the Boundary Operators

The operators defined in the previous section have a number of important properties that will be used extensively in the development of solutions of our boundary value problems.

2.7 Theorem. V_0, W_0, W_0^*, and N_0 satisfy the composition formulas

$$W_0V_0 = V_0W_0^*, \quad N_0V_0 = W_0^{*2} - \tfrac{1}{4}I \quad \text{on } C^{0,\alpha}(\partial S), \tag{2.33}$$

$$N_0W_0 = W_0^*N_0, \quad V_0N_0 = W_0^2 - \tfrac{1}{4}I \quad \text{on } C^{1,\alpha}(\partial S). \tag{2.34}$$

Proof. Let

$$u = \mathscr{V}^+\varphi - \mathscr{W}^+\psi, \tag{2.35}$$

where $\varphi \in C^{0,\alpha}(\partial S)$ and $\psi \in C^{1,\alpha}(\partial S)$ are arbitrary. By Theorem 2.5(i), $(Zu)(x) = 0$ for $x \in S^+$. Using (2.32) and restricting u to ∂S, we get

$$u|_{\partial S} = (\mathscr{V}^+\varphi)|_{\partial S} - (\mathscr{W}^+\psi)|_{\partial S},$$

or, setting $u|_{\partial S} = \alpha$,

$$\alpha = V_0\varphi - \left(W_0 - \tfrac{1}{2}I\right)\psi. \tag{2.36}$$

Applying the boundary stress operator T in (2.35) and using (2.26) and (2.31), we arrive at

$$Tu = T(\mathscr{V}^+\varphi) - T(\mathscr{W}^+\psi);$$

that is, with $Tu = \beta$,

$$\beta = \left(W_0^* - \tfrac{1}{2}I\right)\varphi - N_0\psi. \tag{2.37}$$

Next, we write the Somigliana representation formula (2.29) restricted to the boundary ∂S:

$$(\mathscr{V}^+(Tu))|_{\partial S} - (\mathscr{W}^+(u|_{\partial S}))|_{\partial S} = u|_{\partial S},$$

which is the same as

$$V_0(Tu) - \left(W_0 - \tfrac{1}{2}I\right)u|_{\partial S} = u|_{\partial S},$$

or

$$V_0\beta - \left(W_0 - \tfrac{1}{2}I\right)\alpha = \alpha. \tag{2.38}$$

Finally, applying T to (2.29) and using (2.26) and (2.31), we have

$$T(\mathscr{V}^+(Tu)) - T(\mathscr{W}^+(u|_{\partial S})) = Tu,$$

or

$$\left(W_0^* + \tfrac{1}{2}I\right)\beta - N_0\alpha = \beta. \tag{2.39}$$

System (2.36), (2.37) can be written as

$$\begin{pmatrix} W_0^* + \frac{1}{2}I & -N_0 \\ V_0 & -\left(W_0 - \frac{1}{2}I\right) \end{pmatrix} \begin{pmatrix} \varphi \\ \psi \end{pmatrix} = \begin{pmatrix} \beta \\ \alpha \end{pmatrix}. \tag{2.40}$$

Also, system (2.38), (2.39) takes the form

$$\begin{pmatrix} W_0^* + \frac{1}{2}I & -N_0 \\ V_0 & -\left(W_0 - \frac{1}{2}I\right) \end{pmatrix} \begin{pmatrix} \beta \\ \alpha \end{pmatrix} = \begin{pmatrix} \beta \\ \alpha \end{pmatrix}. \tag{2.41}$$

Replacing $(\beta, \alpha)^{\mathrm{T}}$ from (2.40) in (2.41), we now deduce that

$$\begin{pmatrix} W_0^* + \frac{1}{2}I & -N_0 \\ V_0 & -\left(W_0 - \frac{1}{2}I\right) \end{pmatrix}^2 \begin{pmatrix} \varphi \\ \psi \end{pmatrix} = \begin{pmatrix} W_0^* + \frac{1}{2}I & -N_0 \\ V_0 & -\left(W_0 - \frac{1}{2}I\right) \end{pmatrix} \begin{pmatrix} \varphi \\ \psi \end{pmatrix},$$

and, since the matrix on the left-hand side expands to

$$\begin{pmatrix} \left(W_0 * + \frac{1}{2}I\right)^2 - N_0 V_0 & -\left(W_0^* + \frac{1}{2}I\right)N_0 + N_0\left(W_0 - \frac{1}{2}I\right) \\ V_0\left(W_0^* + \frac{1}{2}I\right) - \left(W_0 - \frac{1}{2}I\right)V_0 & -V_0 N_0 + \left(W_0 - \frac{1}{2}I\right)^2 \end{pmatrix},$$

the usual identification of elements and the arbitrariness of φ and ψ in their corresponding spaces lead to (2.33) and (2.34). $\qquad\square$

2.8 Theorem. $W_0 \pm \frac{1}{2}I$ and $W_0^* \pm \frac{1}{2}I$ are operators of index zero.

The proof of this theorem is very similar to that of Theorem 2.5(v) and of the corresponding assertion in [6], where the terms with the strongest singularity in the kernels of W_0 and W_0^* are the same as here, up to a constant factor. Since the index [6] is homogeneous of order 0 in the numerical coefficient of $\partial_s \ln|x - y|$, we conclude that the index of our operators W_0 and W_0^* is the same as that of their counterparts in [6], namely 0.

A useful tool from functional analysis that will enable us to solve our boundary value problems is the Fredholm alternative, which can be stated using Definitions 1.72 and 1.73 from [7].

2.9 Definition. A *dual system* (X,Y) is a pair of normed spaces X and Y with a non-degenerate bilinear form $(\cdot,\cdot) : X \times Y \to \mathbb{C}$.

2.10 Definition. Let (X,Y) be a dual system with bilinear form $(\cdot,\cdot)$, let $K : X \to X$ be an operator with a (unique) adjoint $K^* : Y \to Y$, let $\omega \in \mathbb{C}$, $\omega \neq 0$, and consider the equations

$$(K - \omega I)\varphi = f, \quad f \in X, \tag{K}$$

$$(K^* - \bar{\omega}I)\varphi = g, \quad g \in X, \tag{K*}$$

together with their homogeneous versions (K_0) and (K_0^*). We say that the *Fredholm alternative* holds for these equations in (X,Y) if either

(i) (K_0) has only the zero solution, in which case so does (K_0^*), and (K) and (K*) have unique solutions for any $f \in X$ and any $g \in Y$; or

(ii) (K_0) and (K_0^*) have the same finite number of linearly independent solutions

$$\{\varphi_1, \varphi_2, \cdots, \varphi_n\} \quad \text{and} \quad \{\psi_1, \psi_2, \cdots, \psi_n\},$$

respectively, and (K) and (K*) are solvable if and only if

$$(f, \psi_i) = 0 \quad \text{and} \quad (g, \varphi_i) = 0, \quad i = 1, 2, \cdots, n.$$

2.11 Corollary. *The Fredholm alternative holds for the pairs of equations*

$$\left(W_0 - \tfrac{1}{2}I\right)\varphi = f, \quad \left(W_0^* - \tfrac{1}{2}I\right)\psi = g,$$
$$\left(W_0 + \tfrac{1}{2}I\right)\varphi = f, \quad \left(W_0^* + \tfrac{1}{2}I\right)\psi = g$$

in the dual system $(C^{0,\alpha}(\partial S), C^{0,\alpha}(\partial S))$, $\alpha \in (0, 1)$, *with the bilinear form*

$$(\varphi, \psi) = \int_{\partial S} \varphi^T \psi \, ds.$$

2.12 Theorem. *The null spaces of* $W_0 - \tfrac{1}{2}I$, $W_0^* - \tfrac{1}{2}I$, $W_0 + \tfrac{1}{2}I$, $W_0^* + \tfrac{1}{2}I$, N_0, *and* V_0 *consist of the zero vector alone.*

Proof. If $\varphi \in C^{0,\alpha}(\partial S)$ is in the null space of $W_0^* - \tfrac{1}{2}I$, then, by (2.26),

$$\left(W_0^* - \tfrac{1}{2}I\right)\varphi = T(\mathscr{V}^-\varphi) = 0$$

and, by (2.27), $Z(\mathscr{V}^-\varphi) = 0$; therefore,

$$\mathscr{V}^-\varphi = 0$$

as the unique solution of the homogeneous problem (N^-). By (2.25) and (2.32), $\mathscr{V}^-\varphi = 0$ implies that

$$(\mathscr{V}^+\varphi)|_{\partial S} = (\mathscr{V}^-\varphi)|_{\partial S} = 0$$

and, by (2.27), $Z(\mathscr{V}^+\varphi) = 0$. Consequently, $\mathscr{V}^+\varphi = 0$ as the unique solution of the homogeneous problem (D^+). Since

$$\mathscr{V}^+\varphi = 0, \quad \mathscr{V}^-\varphi = 0,$$

from Remark 2.6(iii) it follows that $\varphi = 0$.

If $\varphi \in C^{0,\alpha}(\partial S)$ is in the null space of $W_0^* + \tfrac{1}{2}I$, then, by (2.26),

$$\left(W_0^* + \tfrac{1}{2}I\right)\varphi = 0$$

is equivalent to $T(\mathscr{V}^+\varphi) = 0$ and, by (2.27), $Z(\mathscr{V}^+\varphi) = 0$; therefore, $\mathscr{V}^+\varphi = 0$ as the unique solution of the homogeneous problem (N^+). By (2.25) and (2.32), $\mathscr{V}^+\varphi = 0$ implies that

$$(\mathscr{V}^-\varphi)|_{\partial S} = (\mathscr{V}^+\varphi)|_{\partial S} = 0$$

and, by (2.27), $Z(\mathscr{V}^-\varphi) = 0$. Therefore, since

$$(\mathscr{V}^-\varphi)|_{\partial S} = 0, \quad \mathscr{V}^-\varphi \in \mathscr{A},$$

$\mathscr{V}^-\varphi = 0$ as the unique solution of the homogeneous problem (D$^-$). Given that

$$\mathscr{V}^+\varphi = \mathscr{V}^-\varphi = 0,$$

from Remark 2.6(iii) it follows that $\varphi = 0$.

Corollary 2.11 and the Fredholm alternative imply that the null space of $W_0 + \frac{1}{2}I$ and that of $W_0 - \frac{1}{2}I$ also consist only of the zero vector.

If $N_0\varphi = 0$, then, by (2.34),

$$0 = V_0(N_0\varphi) = (V_0 N_0)\varphi$$
$$= \left(W_0^2 - \frac{1}{4}I\right)\varphi = \left(W_0 - \frac{1}{2}I\right)\left(W_0 + \frac{1}{2}I\right)\varphi,$$

so, given the structure of the null spaces of $W_0 - \frac{1}{2}I$ and $W_0 + \frac{1}{2}I$, we conclude that $\varphi = 0$.

By (2.32), $V_0\varphi = 0$ implies that

$$(\mathscr{V}^+\varphi)|_{\partial S} = 0.$$

By (2.27), $Z(\mathscr{V}^+\varphi) = 0$; hence, by Theorem 1.9, $\mathscr{V}^+\varphi = 0$ as the unique solution of the homogeneous problem (D$^+$). Also, by (2.32), $V_0\varphi = 0$ implies that

$$(\mathscr{V}^-\varphi)|_{\partial S} = 0.$$

By (2.27), $Z(\mathscr{V}^-\varphi) = 0$; therefore, by Theorem 1.9, since

$$(\mathscr{V}^-\varphi)|_{\partial S} = 0, \quad \mathscr{V}^-\varphi \in \mathscr{A},$$

$\mathscr{V}^-\varphi = 0$ as the unique solution of the homogeneous problem (D$^-$). The equalities $\mathscr{V}^+\varphi = \mathscr{V}^-\varphi = 0$ and Remark 2.6(iii) now imply that $\varphi = 0$. $\qquad\square$

Chapter 3
Existence of Solutions

In this chapter, we describe and apply two boundary integral equation methods to solve the fundamental boundary value problems formulated in Sect. 1.2.

3.1 The Classical Indirect Method

For the interior and exterior Dirichlet, Neumann, and Robin problems $(D^\pm)$, $(N^\pm)$, and $(R^\pm)$, the solutions are sought in the form

$$
\begin{aligned}
u &= \mathscr{W}^+\varphi && \text{for } (D^+), \\
u &= \mathscr{W}^-\varphi && \text{for } (D^-), \\
u &= \mathscr{V}^+\psi && \text{for } (N^+), \\
u &= \mathscr{V}^-\psi && \text{for } (N^-), \\
u &= \mathscr{V}^+\varphi && \text{for } (R^+), \\
u &= \mathscr{V}^-\varphi && \text{for } (R^-).
\end{aligned}
$$

From the properties of $\mathscr{W}^\pm$ and $\mathscr{V}^\pm$ listed in Theorem 2.5 and the boundary conditions of each of these problems, the unknown density functions φ and ψ must satisfy, respectively, the boundary integral equations

$$
\begin{aligned}
\left(W_0 - \tfrac{1}{2}I\right)\varphi &= \mathscr{P}, && (\mathscr{D}_C^+) \\
\left(W_0 + \tfrac{1}{2}I\right)\varphi &= \mathscr{R}, && (\mathscr{D}_C^-) \\
\left(W_0^* + \tfrac{1}{2}I\right)\psi &= \mathscr{Q}, && (\mathscr{N}_C^+) \\
\left(W_0^* - \tfrac{1}{2}I\right)\psi &= \mathscr{S}, && (\mathscr{N}_C^-) \\
\left(W_0^* + \tfrac{1}{2}I\right)\varphi + \sigma(V_0\varphi) &= \mathscr{K}, && (\mathscr{R}_C^+) \\
\left(W_0^* - \tfrac{1}{2}I\right)\varphi - \sigma(V_0\varphi) &= \mathscr{L}. && (\mathscr{R}_C^-)
\end{aligned}
$$

© Springer International Publishing Switzerland 2016
C. Constanda et al., *Boundary Integral Equation Methods and Numerical Solutions*,
Developments in Mathematics 35, DOI 10.1007/978-3-319-26309-0_3

3.1.1 The Dirichlet Problems

3.1 Theorem. (i) $(\mathscr{D}_C^+)$ *has a unique solution* $\varphi \in C^{1,\alpha}(\partial S)$ *for any* $\mathscr{P} \in C^{1,\alpha}(\partial S)$, $\alpha \in (0,1)$. *Then* (D^+) *has the unique solution*

$$u = \mathscr{W}^+\varphi. \tag{3.1}$$

(ii) $(\mathscr{D}_C^-)$ *has a unique solution* $\varphi \in C^{1,\alpha}(\partial S)$ *for any* $\mathscr{R} \in C^{1,\alpha}(\partial S)$, $\alpha \in (0,1)$. *Then* (D^-) *has the unique solution*

$$u = \mathscr{W}^-\varphi. \tag{3.2}$$

Proof. (i) By Theorem 2.12, the null space of $W_0 - \frac{1}{2}I$ contains only the zero vector, so, by the Fredholm alternative, $(\mathscr{D}_C^+)$ has a unique solution $\varphi \in C^{1,\alpha}(\partial S)$.

Since u defined by (3.1) satisfies $Zu = 0$ and

$$u|_{\partial S} = \mathscr{W}_0^+\varphi = \left(W_0 - \tfrac{1}{2}I\right)\varphi = \mathscr{P},$$

we conclude that this function is the unique solution of (D^+).

(ii) The proof of this assertion is a mirror image of that of (i). Additionally, here we must mention that, by Theorem 2.4, u defined by (3.2) also satisfies the far-field condition $u \in \mathscr{A}$.   □

3.1.2 The Neumann Problems

3.2 Theorem. (i) $(\mathscr{N}_C^+)$ *has a unique solution* $\psi \in C^{1,\alpha}(\partial S)$ *for any* $\mathscr{Q} \in C^{0,\alpha}(\partial S)$, $\alpha \in (0,1)$. *Then* (N^+) *has the unique solution*

$$u = \mathscr{V}^+\psi. \tag{3.3}$$

(ii) $(\mathscr{N}_C^-)$ *has a unique solution* $\psi \in C^{0,\alpha}(\partial S)$ *for any* $\mathscr{S} \in C^{0,\alpha}(\partial S)$, $\alpha \in (0,1)$. *Then* (N^-) *has the unique solution*

$$u = \mathscr{V}^-\psi. \tag{3.4}$$

Proof. (i) By Theorem 2.12, the null space of $W_0^* + \frac{1}{2}I$ contains only the zero vector, so the Fredholm alternative implies that $(\mathscr{N}_C^+)$ has a unique solution $\psi \in C^{1,\alpha}(\partial S)$.

Since u defined by (3.3) satisfies $Zu = 0$ and

$$Tu = T(\mathscr{V}^+\psi) = \left(W_0^* + \tfrac{1}{2}I\right)\psi = \mathscr{Q},$$

we conclude that this function is the unique solution of (N^+).

(ii) The proof of this part is similar to that of (i), with the added mention that, according to Theorem 2.4, u defined by (3.4) also satisfies $u \in \mathscr{A}$.   □

3.1.3 The Robin Problems

3.3 Theorem. (i) $(\mathscr{R}_C^+)$ *has a unique solution* $\varphi \in C^{0,\alpha}(\partial S)$ *for any* $\mathscr{K} \in C^{0,\alpha}(\partial S)$ *and any* $\sigma \in C^{0,\alpha}(\partial S)$, $\alpha \in (0,1)$, *Then* (R^+) *has the unique solution*

$$u = \mathscr{V}^+ \varphi. \tag{3.5}$$

(ii) $(\mathscr{R}_C^-)$ *has a unique solution* $\varphi \in C^{0,\alpha}(\partial S)$ *for any* $\mathscr{L} \in C^{0,\alpha}(\partial S)$ *and any* $\sigma \in C^{0,\alpha}(\partial S)$, $\alpha \in (0,1)$. *Then* (R^-) *has the unique solution*

$$u = \mathscr{V}^- \varphi. \tag{3.6}$$

Proof. (i) Let μ be a solution of the homogeneous equation $(\mathscr{R}_C^+)$; that is,

$$\left(W_0^* + \tfrac{1}{2}I\right)\mu + \sigma(V_0\mu) = 0,$$

which, according to Theorem 2.5(v), means that

$$T(\mathscr{V}^+\mu) + \sigma(\mathscr{V}_0^+\mu) = 0.$$

Hence, $u = \mathscr{V}^+\mu$ is a solution of the homogeneous problem (R^+), so, by Theorem 1.9, $\mathscr{V}^+\mu = 0$. Then, by Theorem 2.5(ii),

$$\mathscr{V}_0^-\mu = \mathscr{V}_0^+\mu = 0.$$

Since $Z(\mathscr{V}^-\mu) = 0$ and, by Theorem 2.4, $\mathscr{V}^-\mu \in \mathscr{A}$, we conclude that $\mathscr{V}^-\mu$ is a solution of (R^-), so, by Theorem 1.9, $\mathscr{V}^-\mu = 0$. Remark 2.6(iii) now implies that $\mu = 0$. Consequently, by the Fredholm alternative, $(\mathscr{R}_C^+)$ has a unique solution $\varphi \in C^{0,\alpha}(\partial S)$.

Given that u defined by (3.5) satisfies $Zu = 0$ and

$$Tu + \sigma u|_{\partial S} = T(\mathscr{V}^+\varphi) + \sigma(\mathscr{V}_0^+\varphi) = \left(W_0^* + \tfrac{1}{2}I\right)\varphi + \sigma(V_0\varphi) = \mathscr{K},$$

this function is the unique solution of (R^+).

(ii) The proof of this assertion is similar to that of (i), with the added observation that u defined by (3.6) also satisfies the far-field condition $u \in \mathscr{A}$. $\qquad\square$

3.2 The Direct Method

In this method, we use the Somigliana representation formulas to set up appropriate boundary integral equations for $(\mathrm{D}^\pm)$, $(\mathrm{N}^\pm)$, and $(\mathrm{R}^\pm)$.

For the interior problems, the restriction of the first equation (2.39) to ∂S yields

$$V_0(Tu) - \left(W_0 + \tfrac{1}{2}I\right)(u|_{\partial S}) = 0.$$

In (D$^+$), $u|_{\partial S} = \mathscr{P}$ is known and $Tu = \varphi$ is unknown; in (N$^+$), $Tu = \mathscr{Q}$ is known and $u|_{\partial S} = \psi$ is unknown. Hence, the corresponding boundary integral equations are

$$V_0\varphi = \left(W_0 + \tfrac{1}{2}I\right)\mathscr{P}, \qquad\qquad (\mathscr{D}_D^+)$$

$$\left(W_0 + \tfrac{1}{2}I\right)\psi = V_0\mathscr{Q}. \qquad\qquad (\mathscr{N}_D^+)$$

For the exterior problems, the first equation (2.40) restricted to ∂S is written as

$$-V_0(Tu) + \left(W_0 - \tfrac{1}{2}I\right)(u|_{\partial S}) = 0.$$

In (D$^-$), $u|_{\partial S} = \mathscr{R}$ is known and $Tu = \varphi$ is unknown; in (N$^-$), $Tu = \mathscr{S}$ is known and $u|_{\partial S} = \psi$ is unknown. Hence, the corresponding boundary integral equations are

$$V_0\varphi = \left(W_0 - \tfrac{1}{2}I\right)\mathscr{R}, \qquad\qquad (\mathscr{D}_D^-)$$

$$\left(W_0 - \tfrac{1}{2}I\right)\psi = V_0\mathscr{S}. \qquad\qquad (\mathscr{N}_D^-)$$

Both $(\mathscr{D}_D^+)$ and $(\mathscr{D}_D^-)$ are equations of the first kind, whereas $(\mathscr{N}_D^+)$ and $(\mathscr{N}_D^-)$ are equations of the second kind.

For (R$^+$), we start with the boundary condition written in the form

$$Tu = -\sigma(u|_{\partial S}) + \mathscr{K},$$

which, on substitution in the first equation (2.29), leads to

$$u = \mathscr{V}^+(-\sigma(u|_{\partial S}) + \mathscr{K}) - \mathscr{W}^+(u|_{\partial S}).$$

Restricting this equation to ∂S, setting

$$\psi = u|_{\partial S},$$

and using (2.32), we find that

$$\psi = -V_0(\sigma\psi) + \left(W_0 - \tfrac{1}{2}I\right)\psi - V_0(\mathscr{K}),$$

or

$$V_0(\sigma\psi) + \left(W_0 + \tfrac{1}{2}I\right)\psi = V_0\mathscr{K}. \qquad\qquad (\mathscr{R}_D^+)$$

For (R$^-$), writing the boundary condition in the form

$$Tu = \sigma(u|_{\partial S}) + \mathscr{L}$$

and using a similar argument, we arrive that

$$V_0(\sigma\psi) - \left(W_0 - \tfrac{1}{2}I\right)\psi = -V_0\mathscr{L}. \qquad\qquad (\mathscr{R}_D^-)$$

3.2.1 The Dirichlet Problems

3.4 Theorem. (i) $(\mathscr{D}_D^+)$ *has a unique solution* $\varphi \in C^{0,\alpha}(\partial S)$ *for any* $\mathscr{P} \in C^{1,\alpha}(\partial S)$, $\alpha \in (0,1)$. *Then* (D^+) *has the unique solution*

$$u = \mathscr{V}^+\varphi - \mathscr{W}^+\mathscr{P}. \tag{3.7}$$

(ii) $(\mathscr{D}_D^-)$ *has a unique solution* $\varphi \in C^{0,\alpha}(\partial S)$ *for any* $\mathscr{R} \in C^{1,\alpha}(\partial S)$, $\alpha \in (0,1)$. *Then* (D^-) *has the unique solution*

$$u = -\mathscr{V}^-\varphi + \mathscr{W}^-\mathscr{R}. \tag{3.8}$$

Proof. Applying N_0 to $(\mathscr{D}_D^+)$, we get

$$N_0(V_0\varphi) = N_0\big((W_0 + \tfrac{1}{2}I)\,\mathscr{P}\big).$$

By Theorem 2.7,

$$N_0(V_0\varphi) = \big(W_0^{*2} - \tfrac{1}{4}I\big)\varphi$$

and

$$N_0\big[(W_0 + \tfrac{1}{2}I)\,\mathscr{P}\big] = W^*(N_0\mathscr{P}) + \tfrac{1}{2}N_0\mathscr{P} = (W^* + \tfrac{1}{2}I)(N_0\mathscr{P});$$

hence,

$$\big(W_0^{*2} - \tfrac{1}{4}I\big)\varphi = \big(W^* + \tfrac{1}{2}I\big)(N_0\mathscr{P}),$$

or

$$\big(W_0^* + \tfrac{1}{2}I\big)\big[\big(W_0^* - \tfrac{1}{2}I\big)\varphi - N_0\mathscr{P}\big] = 0.$$

By Theorem 2.12, the null space of $W_0^* + \tfrac{1}{2}I$ consists only of the zero vector; therefore,

$$\big(W_0^* - \tfrac{1}{2}I\big)\varphi - N_0\mathscr{P} = 0.$$

Since, by the same theorem, the null space of $W_0^* - \tfrac{1}{2}I$ also consists only of the zero vector, from the Fredholm alternative it follows that the above equation has a unique solution φ, which is also the unique solution of $(\mathscr{D}_D^+)$.

Given that u defined by (3.7) satisfies

$$Zu = 0 \quad \text{in } S^+$$

and, in view of $(\mathscr{D}_D^+)$,

$$\begin{aligned}
u|_{\partial S} &= (\mathscr{V}^+\varphi)|_{\partial S} - (\mathscr{W}^+\mathscr{P})|_{\partial S} \\
&= V_0\varphi - \big(W_0 - \tfrac{1}{2}I\big)\mathscr{P} \\
&= V_0\varphi - \big(W_0 + \tfrac{1}{2}I\big)\mathscr{P} + \mathscr{P} = \mathscr{P},
\end{aligned}$$

we conclude that (3.7) is the unique solution of (D^+).

(ii) The proof of this assertion is similar to that of (i), with the additional mention that, by Theorem 2.4, u defined by (3.8) also satisfies $u \in \mathscr{A}$. $\qquad \square$

3.2.2 The Neumann Problems

3.5 Theorem. (i) $(\mathcal{N}_D^+)$ *has a unique solution* $\psi \in C^{1,\alpha}(\partial S)$ *for any* $\mathcal{Q} \in C^{0,\alpha}(\partial S)$, $\alpha \in (0,1)$. *Then* (N^+) *has the unique solution*

$$u = \mathcal{V}^+ \mathcal{Q} - \mathcal{W}^+ \psi. \tag{3.9}$$

(ii) $(\mathcal{N}_D^-)$ *has a unique solution* $\psi \in C^{1,\alpha}(\partial S)$ *for any* $\mathcal{S} \in C^{0,\alpha}(\partial S)$, $\alpha \in (0,1)$. *Then* (N^-) *has the unique solution*

$$u = -\mathcal{V}^- \mathcal{Q} + \mathcal{W}^- \psi. \tag{3.10}$$

Proof. (i) By Theorem 2.12, the null space of $W_0 + \frac{1}{2}I$ consists only of the zero vector, so, by the Fredholm alternative, $(\mathcal{N}_D^+)$ has a unique solution $\psi \in C^{1,\alpha}(\partial S)$.

Clearly, u defined by (3.9) satisfies

$$Zu = 0 \quad \text{in } S^+.$$

To show that it also satisfies the boundary condition $Tu = \mathcal{Q}$, we apply T to both sides in (3.9):

$$Tu = \left(W_0^* + \tfrac{1}{2}I\right)\mathcal{Q} - N_0\psi. \tag{3.11}$$

Let

$$\mathcal{H} = \left(W_0^* + \tfrac{1}{2}I\right)\mathcal{Q} - N_0\psi - \mathcal{Q} = \left(W_0^* - \tfrac{1}{2}I\right)\mathcal{Q} - N_0\psi.$$

Operating on both sides above with V_0 and using (2.34), we see that

$$\begin{aligned}
V_0\mathcal{H} &= V_0\left[\left(W_0^* - \tfrac{1}{2}I\right)\mathcal{Q} - N_0\psi\right] \\
&= (V_0 W_0^*)\mathcal{Q} - \tfrac{1}{2}V_0\mathcal{Q} - \left(W_0^2 - \tfrac{1}{4}I\right)\psi \\
&= (W_0 V_0)\mathcal{Q} - \tfrac{1}{2}V_0\mathcal{Q} - \left(W_0^2 - \tfrac{1}{4}I\right)\psi \\
&= \left(W_0 - \tfrac{1}{2}I\right)V_0\mathcal{Q} - \left(W_0^2 - \tfrac{1}{4}I\right)\psi \\
&= \left(W_0 - \tfrac{1}{2}I\right)\left[V_0\mathcal{Q} - \left(W_0 - \tfrac{1}{2}I\right)\psi\right] = 0.
\end{aligned}$$

Therefore, by Theorem 2.12, $\mathcal{H} = 0$, which leads to

$$\left(W_0^* + \tfrac{1}{2}I\right)\mathcal{Q} - N_0\psi - \mathcal{Q} = 0,$$

or

$$\left(W_0^* + \tfrac{1}{2}I\right)\mathcal{Q} - N_0\psi = \mathcal{Q};$$

hence, in view of (3.11),

$$Tu = \mathcal{Q}.$$

(ii) The proof of this part is similar to that of part (i), with the additional observation that, by Theorem 2.4, $u \in \mathcal{A}$. $\qquad\square$

3.2.3 The Robin Problems

3.6 Theorem. (i) $(\mathscr{R}_D^+)$ *has a unique solution* $\psi \in C^{1,\alpha}(\partial S)$ *for any* $\mathscr{K} \in C^{0,\alpha}(\partial S)$ *and any* $\sigma \in C^{0,\alpha}(\partial S)$, $\alpha \in (0,1)$. *Then* (R^+) *has the unique solution*

$$u = -\mathscr{V}^+(\sigma\psi) - \mathscr{W}^+\psi + \mathscr{V}^+\mathscr{K}. \tag{3.12}$$

(ii) $(\mathscr{R}_D^-)$ *has a unique solution* $\psi \in C^{1,\alpha}(\partial S)$ *for any* $\mathscr{L} \in C^{0,\alpha}(\partial S)$ *and any* $\sigma \in C^{0,\alpha}(\partial S)$, $\alpha \in (0,1)$. *Then* (R^-) *has the unique solution*

$$u = -\mathscr{V}^-(\sigma\psi) + \mathscr{W}^-\psi - \mathscr{V}^-\mathscr{K}. \tag{3.13}$$

Proof. (i) The homogeneous version of $(\mathscr{R}_D^+)$, namely

$$V_0(\sigma\psi) + \left(W_0 + \tfrac{1}{2}I\right)\psi = 0,$$

can be rewritten as

$$V_0(\sigma\psi) + \left(W_0 - \tfrac{1}{2}I\right)\psi = -\psi. \tag{3.14}$$

Let

$$\mathscr{U}^- = \mathscr{V}^-(\sigma\psi) + \mathscr{W}^-(\psi).$$

Then $Z\mathscr{U}^- = 0$ in S^- and

$$\mathscr{U}^-|_{\partial S} = V_0(\sigma\psi) + \left(W_0 + \tfrac{1}{2}I\right)\psi,$$

which means that $\mathscr{U}^-$ is a solution of the homogeneous problem (D^-), so, by Theorem 1.9,

$$\mathscr{U}^- = 0.$$

Next, applying T to $\mathscr{U}^-$ and using (2.26) and (2.32), we obtain

$$T(\mathscr{V}^-(\sigma\psi)) + T(\mathscr{W}^-(\psi)) = 0,$$

so

$$\left(W_0^* - \tfrac{1}{2}I\right)(\sigma\psi) + N_0\psi = 0,$$

or

$$\left(W_0^* + \tfrac{1}{2}I\right)(\sigma\psi) + N_0\psi = \sigma\psi. \tag{3.15}$$

Let

$$\mathscr{U}^+ = \mathscr{V}^+(\sigma\psi) + \mathscr{W}^+(\psi).$$

Then $Z\mathscr{U}^+ = 0$ in S^+ and, by (2.26), (2.31), (3.15), and (3.14),

$$
\begin{aligned}
T\mathscr{U}^+ + \sigma\mathscr{U}^+|_{\partial S} &= T[\mathscr{V}^+(\sigma\psi) + \mathscr{W}^+\psi] + \sigma[\mathscr{V}^+(\sigma\psi) + \mathscr{W}^+(\psi)]|_{\partial S} \\
&= \left(W_0^* + \tfrac{1}{2}I\right)(\sigma\psi) + N_0\psi + \sigma\left[V_0(\sigma\psi) + \left(W_0 - \tfrac{1}{2}I\right)\psi\right] \\
&= \sigma\psi - \sigma\psi = 0.
\end{aligned}
$$

Hence, $\mathscr{U}^+$ is a solution to the homogeneous problem (R^+), and from Theorem 1.9 it follows that $\mathscr{U}^+ = 0$, which means that the homogeneous equation $(\mathscr{R}_{\mathrm{D}}^+)$ has a unique solution. Then, by the Fredholm alternative, so does $(\mathscr{R}_{\mathrm{D}}^+)$.

It is obvious that u defined by (3.12) satisfies

$$Zu = 0 \quad \text{in } S^+.$$

Restricting (3.12) to ∂S and using $(\mathscr{R}_{\mathrm{D}}^+)$, (2.26), (2.31), and (2.32), we see that

$$
\begin{aligned}
Tu &= -\left(W_0^* + \tfrac{1}{2}I\right)(\sigma\psi) - N_0\psi + \left(W_0^* + \tfrac{1}{2}I\right)\mathscr{K} \\
&= \left(W_0^* + \tfrac{1}{2}I\right)(\mathscr{K} - \sigma\psi) - N_0\psi,
\end{aligned}
$$

so

$$
\begin{aligned}
Tu + \sigma u|_{\partial S} &= \left(W_0^* + \tfrac{1}{2}I\right)(\mathscr{K} - \sigma\psi) - N_0\psi \\
&\quad + \sigma\left[V_0(-\sigma\psi) - \left(W_0 - \tfrac{1}{2}I\right)\psi + V_0\mathscr{K}\right] \\
&= \left(W_0^* + \tfrac{1}{2}I\right)(\mathscr{K} - \sigma\psi) - N_0\psi \\
&\quad + \sigma\left[V_0(\mathscr{K} - \sigma\psi) - \left(W_0 + \tfrac{1}{2}I\right)\psi + \psi\right] \\
&= \left(W_0^* + \tfrac{1}{2}I\right)(\mathscr{K} - \sigma\psi) - N_0\psi + \sigma\psi. \tag{3.16}
\end{aligned}
$$

Let

$$
\begin{aligned}
\Psi &= \left(W_0^* + \tfrac{1}{2}I\right)(\mathscr{K} - \sigma\psi) - N_0\psi + \sigma\psi - \mathscr{K} \\
&= \left(W_0^* - \tfrac{1}{2}I\right)(\mathscr{K} - \sigma\psi) - N_0\psi. \tag{3.17}
\end{aligned}
$$

By $(\mathscr{R}_{\mathrm{D}}^+)$, (2.33), and (2.34),

$$
\begin{aligned}
V_0\Psi &= V_0\left[\left(W_0^* - \tfrac{1}{2}I\right)(\mathscr{K} - \sigma\psi) - N_0\psi\right] \\
&= V_0\left(W_0^* - \tfrac{1}{2}I\right)(\mathscr{K} - \sigma\psi) - V_0(N_0\psi) \\
&= \left(W_0 - \tfrac{1}{2}I\right)V_0(\mathscr{K} - \sigma\psi) - \left(W_0^2 - \tfrac{1}{4}I\right)\psi \\
&= \left(W_0 - \tfrac{1}{2}I\right)\left[V_0(\mathscr{K} - \sigma\psi) - \left(W_0 + \tfrac{1}{2}I\right)\psi\right] = 0.
\end{aligned}
$$

According to (2.32) and Theorem 2.12, $V_0\Psi = 0$ implies that $\Psi = 0$; therefore, by (3.17) and (3.16),

$$\left(W_0^* + \tfrac{1}{2}I\right)(\mathscr{K} - \sigma\psi) - N_0\psi + \sigma\psi = \mathscr{K},$$

or

$$Tu + \sigma u|_{\partial S} = \mathscr{K},$$

which confirms that (3.12) is the unique solution of (R^+).

(ii) The proof of this part is similar to that of part (i), with the additional detail that, by Theorem 2.4, the function defined by (3.13) also satisfies $u \in \mathscr{A}$. □

3.7 Remark. An alternative way to reduce (R^+) to a boundary integral equation is to write the boundary condition in the form

$$u|_{\partial S} = \sigma^{-1}(\mathscr{K} - Tu),$$

which, replaced in the first equality (2.29) restricted to ∂S and combined with (2.32) and the notation

$$\varphi = Tu,$$

leads to

$$V_0 \varphi + \left(W_0 + \tfrac{1}{2}I\right)(\sigma^{-1}\varphi) = \left(W_0 + \tfrac{1}{2}I\right)(\sigma^{-1}\mathscr{K}). \qquad (\mathscr{R}_A^+)$$

A similar equation can be set up for the problem (R^-), and an existence theorem analogous to Theorem 3.6 can be proved.

3.8 Remarks. (i) The existence and uniqueness results established in this chapter remain valid if the boundary ∂S is closed, continuous, and piecewise smooth—for simplicity of the analytic argument, in our case we assume piecewise continuous tangent and curvature—since this allows us to apply the divergence theorem. However, in such domains we expect the solution to lose some degree of smoothness. In a rectangle, for example, this happens as we approach a corner, even if the Dirichlet data are continuous there. Chapter 5 contains several numerical approximations for domains of this type.

(ii) The direct and indirect methods can also be adapted to the case when the boundary conditions are of mixed type—for example, when Dirichlet conditions are prescribed on one part of the boundary and Neumann conditions on the remaining part.

Chapter 4
Software Development

4.1 Programming Environment

All computational, symbolic, numerical, and graphic software used in the rest of the book has been generated with *Mathematica*®, a commercial product of Wolfram Research. The purpose of this chapter is to both document and illustrate how the software has been developed. Many issues specific to the boundary integral methods that form the object of our study require special attention. Below, we summarize these issues and indicate how certain difficulties associated with them can be overcome in the *Mathematica*® environment. We do not intend to give specific detailed code, but rather a general outline of how the code has been assembled.

To simplify the terminology, in what follows the words 'code' and 'coding' refer to programming with *Mathematica*®. Also, the indentation of the displayed formulas has been changed to conform to the TEX output of the software.

4.2 Notation

Mathematica® is at its core a functional language. All functions and their arguments are given a name coded in the format

name[arg1, arg2, ..., argn].

This allows great flexibility in generating functions that accomplish the intended tasks. However, such notation is both cumbersome and, often, difficult to follow, which has prompted us to replace the most commonly used functional calls in our software by suggestive symbolic notation.

4.1 Example. It is very convenient to code *Mathematica*® definitions for the usual infix operators as $\times$, $\circ$, and $\diamond$, respectively, to denote the scalar, functional, and operator compositional 'inner products'. Thus, the composition of a matrix of oper-

© Springer International Publishing Switzerland 2016

C. Constanda et al., *Boundary Integral Equation Methods and Numerical Solutions*,

Developments in Mathematics 35, DOI 10.1007/978-3-319-26309-0_4

ators with a matrix of functions by means of the operator $\diamond$ is written as

$$\begin{pmatrix} f_{1,1}[\#]\& \ f_{1,2}[\#]\& \\ f_{2,1}[\#]\& \ f_{2,2}[\#]\& \end{pmatrix} \diamond \begin{pmatrix} g_1[t] \\ g_2[t] \end{pmatrix} = \begin{pmatrix} f_{1,1}[g_1[t]] + f_{1,2}[g_2[t]] \\ f_{2,1}[g_1[t]] + f_{2,2}[g_2[t]] \end{pmatrix}.$$

Here, the functions are entered in *Mathematica*® notation as pure functions by means of the customary notation

$$f_{i,j}[\#]\&.$$

This type of symbolic notational definitions can be combined with others to build complicated expressions that take on a more traditional mathematical appearance.

4.2 Example. If for the line integral we introduce

$$\text{Notation}\left[\oint_\Gamma f_{\_} \ d\Gamma_{t\_} \quad \Leftrightarrow \quad \text{lineIntegral}\Gamma[f_{\_},t_{\_}]\right],$$

then in our code we can use the familiar form

$$\oint_\Gamma f[t] \, d\Gamma_t$$

instead of its rather obscure equivalent

lineIntegral$\Gamma[f[t],t]$.

4.3 Example. The representation formula (2.18) for $x \in S^+$ quoted later in this chapter is coded as

$$u[x] = \oint_\Gamma D[x,y[t]] \circ (\mathrm{T}u)[y[t]] \times \mathrm{dsdt}[t] \, d\Gamma_t$$

$$- \oint_\Gamma P[x,y[t]] \circ u[y[t]] \times \mathrm{dsdt}[t] \, d\Gamma_t.$$

In what follows, expressions related to code in *Mathematica*® are displayed as much as possible in this suggestive notation. Also, whenever appropriate, specific functional calls will reference equations and formulas developed earlier in the book, to aid the readers in their interpretation of certain important details.

4.4 Remark. The text in Chaps. 4 and 5 contains a mixture of mathematical symbols and *Mathematica*® objects. To distinguish between the two, the former are written in the standard mathematical font used in the first three chapters, and the latter in a combination of fonts according to the conventions embedded in the software. Sometimes, both categories of symbols need to stand side by side in the same formula.

4.3 Coding the Mathematical Model

The system of equations describing the equilibrium state of a thin plate on an elastic foundation was described in Chap. 1. These equations, defined in regions S^+ and S^- with boundary ∂S (see Fig. 1.1), form the linear and homogeneous elliptic system (1.8). In *Mathematica*®, using notational definitions, we enter the differential operator (1.6) as

$$Z_x := \begin{pmatrix} \mu(\partial_{x_1,x_1}\# + \partial_{x_2,x_2}\#) + (\lambda+\mu)\partial_{x_1,x_1}\# - k\#\& & (\lambda+\mu)\partial_{x_1,x_2}\#\& \\ (\lambda+\mu)\partial_{x_1,x_2}\#\& & \mu(\partial_{x_1,x_1}\# + \partial_{x_2,x_2}\#) + (\lambda+\mu)\partial_{x_2,x_2}\# - k\#\& \end{pmatrix},$$

(4.1)

where the subscript x on the left-hand side indicates that the operator Z acts with respect to the point x. This subscript becomes relevant when Z is applied to two-point functions.

Similarly, the coding of the boundary stress operator (1.10) is

$$T_x := \begin{pmatrix} (\lambda+2\mu)v1[x_1,x_2]\partial_{x_1}\# + \mu v2[x_1,x_2]\partial_{x_2}\#\& & \mu v2[x_1,x_2]\partial_{x_1}\# + \lambda v1[x_1,x_2]\partial_{x_2}\#\& \\ \lambda v2[x_1,x_2]\partial_{x_1}\# + \mu v1[x_1,x_2]\partial_{x_2}\#\& & \mu v1[x_1,x_2]\partial_{x_1}\# + (\lambda+2\mu)v2[x_1,x_2]\partial_{x_2}\#\& \end{pmatrix}.$$

(4.2)

The homogeneous system (1.8), now rewritten as

$$Z_x \diamond u = 0, \quad x \in S^+ \cup S^-,$$

together with Dirichlet, Neumann, or Robin boundary conditions prescribed on ∂S constitute a boundary value problem (see Sect. 1.2). In the rest of the book, we focus our attention exclusively on problems in $S^+ \cup \partial S$ and, for brevity, denote S^+ by S.

4.4 Coding of the Matrix of Fundamental Solutions

The symbolic capabilities of *Mathematica*® allow us to compute the explicit form of the matrix of fundamental solutions constructed in Sect. 2.1. As shown below, these solutions are nontrivial, and the software both removes the potential for derivation errors and makes it possible for us to verify the result. In our case, the 2×2 matrix of fundamental solutions D satisfies (see (2.1))

$$Z_x \diamond D[x,y] = Z_x \diamond \begin{pmatrix} D_{1,1}[x,y] & D_{1,2}[x,y] \\ D_{2,1}[x,y] & D_{2,2}[x,y] \end{pmatrix} = -\delta[x,y] \begin{pmatrix} 1 & 0 \\ 0 & 1 \end{pmatrix}.$$

For convenience, we use our own coding definitions to recall the procedure for constructing D. According to (2.2), D is of the form

$$D[x,y] = \text{adjoint}[Z_x] \diamond \begin{pmatrix} t[x,y] & 0 \\ 0 & t[x,y] \end{pmatrix},$$

(4.3)

which, substituted in the equilibrium system, yields

$$Z_x \diamond \text{adjoint}[Z_x] \diamond \begin{pmatrix} t[x,y] & 0 \\ 0 & t[x,y] \end{pmatrix} = \text{Det}[Z_x] \diamond \begin{pmatrix} t[x,y] & 0 \\ 0 & t[x,y] \end{pmatrix}$$

$$= -\delta[x,y] \begin{pmatrix} 1 & 0 \\ 0 & 1 \end{pmatrix}.$$

The matrix of fundamental solutions is therefore constructed from the solution of the equation

$$\text{Det}[Z_x] \diamond t[x,y] = -\delta[x,y]. \tag{4.4}$$

Mathematica® evaluates the symbolic determinant operator $\text{Det}[Z_x]$ as

$$\left(k^2 \#1 - k(\lambda + 3\mu)\partial_{x1,x1}\#1 - k(\lambda + 3\mu)\partial_{x2,x2}\#1 + \lambda\mu\partial_{x1,x1,x1,x1}\#1 \right.$$
$$+ 2\mu^2\partial_{x1,x1,x1,x1}\#1 + 2\lambda\mu\partial_{x1,x1,x2,x2}\#1 + 4\mu^2\partial_{x1,x1,x2,x2}\#1$$
$$\left. + \lambda\mu\partial_{x2,x2,x2,x2}\#1 + 2\mu^2\partial_{x2,x2,x2,x2}\#1 \right) \&,$$

which simplifies to

$$\left(k^2 \mathfrak{J} - k(\lambda + 3\mu)\Delta + \mu(\lambda + 2\mu)\Delta\Delta \right) \&; \tag{4.5}$$

here, $\mathfrak{J}$ denotes the identity operator/matrix since the symbol I, used in the earlier chapters, has a different assignment in *Mathematica*®.

This operator can be formally factored, so the equation for t becomes

$$\mu(\lambda + 2\mu)\left(\Delta - \frac{k}{\mu}\mathfrak{J} \right)\left(\Delta - \frac{k}{\lambda + 2\mu}\mathfrak{J} \right) \diamond t[x,y] = -\delta[x,y].$$

Converting the above equation to polar form, solving it symbolically, and keeping, as indicated in Sect. 2.1, only the terms with singularities at $r = 0$ leads to t being expressed as the combination of modified Bessel functions (see (2.5))

$$t[r] = \frac{1}{2k\pi(\lambda + \mu)}\left(\text{BesselK}\left[0, r\sqrt{\frac{k}{\mu}}\right] - \text{BesselK}\left[0, r\sqrt{\frac{k}{\lambda + 2\mu}}\right] \right).$$

At this time, our code can be used to verify the accuracy of this expression by checking symbolically that (4.4) is satisfied.

All the necessary elements are now available to compute the matrix of fundamental solutions (4.3), where (see (2.6))

adjoint $[Z_x]$

$$= \begin{pmatrix} -k\#1 + \mu\partial_{x1,x1}\#1 + (\lambda + 2\mu)\partial_{x2,x2}\#1\& & -(\lambda + \mu)\partial_{x1,x2}\#1\& \\ -(\lambda + \mu)\partial_{x1,x2}\#1\& & -k\#1 + (\lambda + 2\mu)\partial_{x1,x1}\#1 + \mu\partial_{x2,x2}\#1\& \end{pmatrix}.$$

Given the excessive size of the full expression, we show explicitly only four of the 60 terms that make up the component $D_{1,1}$ of D:

$$D_{1,1}[x1, x2, y1, y2] = \cdots$$

$$+ \frac{\text{BesselK}\left[0, \sqrt{(x1-y1)^2 + (x2-y2)^2}\sqrt{\frac{k}{\mu}}\right]}{2\pi(\lambda+\mu)} + \cdots$$

$$+ \frac{x2y2\mu\, \text{BesselK}\left[0, \sqrt{(x1-y1)^2 + (x2-y2)^2}\sqrt{\frac{k}{\lambda+2\mu}}\right]}{\pi\left((x1-y1)^2 + (x2-y2)^2\right)(\lambda+\mu)(\lambda+2\mu)} + \cdots$$

$$+ \frac{2x2y2\mu\sqrt{\frac{k}{\lambda+2\mu}}\, \text{BesselK}\left[1, \sqrt{(x1-y1)^2 + (x2-y2)^2}\sqrt{\frac{k}{\lambda+2\mu}}\right]}{k\pi\left((x1-y1)^2 + (x2-y2)^2\right)^{3/2}(\lambda+\mu)} + \cdots$$

$$+ \frac{x2y2\lambda\, \text{BesselK}\left[2, \sqrt{(x1-y1)^2 + (x2-y2)^2}\sqrt{\frac{k}{\mu}}\right]}{2\pi\left((x1-y1)^2 + (x2-y2)^2\right)\mu(\lambda+\mu)} + \cdots .$$

It can be partially verified that this expression is correctly computed by using the symbolic capabilities of *Mathematica*® to check that

$$Z_x \diamond D[x, y] = \begin{pmatrix} 0 & 0 \\ 0 & 0 \end{pmatrix}, \quad x \neq y. \tag{4.6}$$

As expected, $D_{1,1}$ and $D_{2,2}$ have logarithmic singularities. *Mathematica*® confirms this: when instructed to expand each of the four components, converted to polar form, in a Taylor series (see (2.13)), the result indicates that

$$D[r] = \begin{pmatrix} O[\text{Log}[r]] & O[1] \\ O[1] & O[\text{Log}[r]] \end{pmatrix}. \tag{4.7}$$

4.5 Remark. Our boundary integrals are of the form

$$\oint_{\partial S} K[x, y] f[y] \, ds[y], \tag{4.8}$$

where the elements of the matrix kernel K may have finite jump discontinuities or a weak or strong singularity at $x = y$. Theoretically, since $y \in \partial S$,

(i) when $x \in S$, (4.8) is a proper definite integral that can be computed without difficulty, however close x is to ∂S;

(ii) when $x \in \partial S$, (4.8) is computed as an improper or a Cauchy principal value integral.

The situation is less clear-cut if (4.8) is evaluated numerically. *Mathematica*® uses floating point arithmetic, so the coordinates of a point are always computed with a certain degree of inaccuracy. This means that a point x lying in S very close to the boundary may be placed numerically on ∂S, or, the other way around, a point y on

∂S may be located numerically inside S in the immediate vicinity of the boundary. Consequently, numerical evaluation of (4.8) may or may not conform to cases (i) and (ii) above. It is therefore necessary for us to examine the behavior of $K[x,y]$ in all possible permutations: both x and y in S, both x and y on ∂S, x in S and y on ∂S, and x on ∂S and y in S.

4.6 Example. Figure 4.1 illustrates graphically how each component of the matrix of fundamental solutions $D[x,y]$ behaves if both x and y are in S. Taking, say, $(x1, x2) = (0.5, 0.5)$ as a test point inside the domain, we notice that $D_{1,1}[x,y]$ and $D_{2,2}[x,y]$ have a logarithmic singularity and that $D_{1,2}[x,y]$ and $D_{2,1}[x,y]$ have a finite jump discontinuity at $(y1, y2) = (0.5, 0.5)$, whose size depends on the direction of the path taken to approach the singular point. These graphs confirm our expectation that the numerical computation of boundary integrals with kernel $D[x,y]$ will not encounter major difficulties when the singular point is in the interior of the domain. However, this is not so when the singularity lies on the boundary.

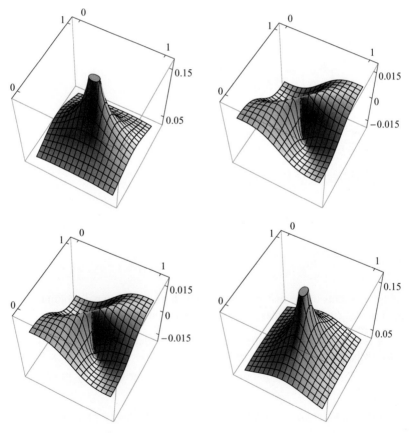

Fig. 4.1 The components of $D[x,y]$, $x,y \in S$.

4.7 Example. Consider again the point $(x1, x2) = (0.5, 0.5)$, but this time for a problem where the boundary is the line $y1 = 0.5$. Figure 4.2 shows the behavior of the four components of $D[x, y]$ when the point $(y1, y2)$ is close to $(0.5, 0.5)$. Their graphs are displayed for the domain to the left of the boundary and truncated to the right of it. The logarithmic singularity of $D_{1,1}[x, y]$ and $D_{2,2}[x, y]$, evidenced by their graphs, becomes an important issue when we use numerical techniques to evaluate boundary integrals of the form (4.8) with $K[x, y] = D[x, y]$. Special care must be taken in such cases to preserve the accuracy of the computed result. We also see that the graphs of $D_{1,2}[x, y]$ and $D_{2,1}[x, y]$ have a finite jump discontinuity at $x = y$. Analytic arguments indicate that this does not happen if the boundary curve has a continuous tangent. For the specific case shown above, the corresponding graphs illustrate the lack of a jump discontinuity along the boundary curve at the point $(y1, y2) = (0.5, 0.5)$. However, unlike the purely analytic treatment, numerical integration is not exact, and if the calculated point $(y1, y2)$ is slightly off the boundary at $y2 = 0.5$, a jump discontinuity occurs, which can produce significant computational errors.

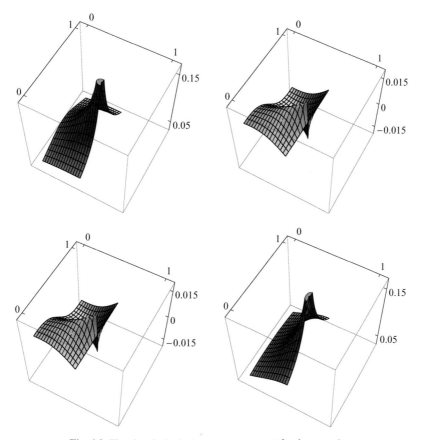

Fig. 4.2 The singularity in the components of $D[x, y]$, $x, y \in \partial S$.

Next, we construct symbolically the 2×2 matrix P, which is used to define the double-layer potential (see (2.9)):

$$P[x,y] := (T_y \diamond D[y,x])^T. \tag{4.9}$$

We do this by direct substitution of the relevant quantities in the above formula, taking care of the necessary switch between x and y. The expression of P is very large, with a LeafCount of 29,183, which exceeds that for D by a factor of 10. In ordinary terms, this means that writing in full the four components of P would produce over 40 pages of printout. This underscores the importance of the symbolic computational abilities of *Mathematica*®.

To better understand the internal structure of P, here we show only a small portion of $P_{1,1}$; specifically, we list a sample of four of the 240 terms that make up this component:

$$P_{1,1}[x1,x2,y1,y2] = \cdots$$

$$+ \frac{3x1^2 y2 \mu \, \text{BesselK}\left[0, \sqrt{(x1-y1)^2+(x2-y2)^2}\sqrt{\frac{k}{\mu}}\right] v2[y1,y2]}{4\pi\left((x1-y1)^2+(x2-y2)^2\right)^2 (\lambda+\mu)} + \cdots$$

$$+ \frac{3x1y1y2\mu^2 \sqrt{\frac{k}{\lambda+2\mu}} \, \text{BesselK}\left[1, \sqrt{(x1-y1)^2+(x2-y2)^2}\sqrt{\frac{k}{\lambda+2\mu}}\right] v2[y1,y2]}{k\pi\left((x1-y1)^2+(x2-y2)^2\right)^{5/2} (\lambda+\mu)} + \cdots$$

$$+ \frac{9x2y2^2\mu \, \text{BesselK}\left[2, \sqrt{(x1-y1)^2+(x2-y2)^2}\sqrt{\frac{k}{\mu}}\right] v2[y1,y2]}{2\pi\left((x1-y1)^2+(x2-y2)^2\right)^2 (\lambda+\mu)} + \cdots$$

$$+ \frac{y2^3\mu^2 \sqrt{\frac{k}{\lambda+2\mu}} \, \text{BesselK}\left[3, \sqrt{(x1-y1)^2+(x2-y2)^2}\sqrt{\frac{k}{\lambda+2\mu}}\right] v2[y1,y2]}{4\pi\left((x1-y1)^2+(x2-y2)^2\right)^{3/2} (\lambda+\mu)(\lambda+2\mu)} + \cdots.$$

Again, it can be partially verified that the above expression of P satisfies the governing system (at all points other than the singularity) by having *Mathematica*® directly evaluate and check that

$$Z_x \diamond P[x,y] = \begin{pmatrix} 0 & 0 \\ 0 & 0 \end{pmatrix}, \quad x \neq y. \tag{4.10}$$

The nature of the singularities of $P[x,y]$ at $x = y$ can be now be studied by instructing *Mathematica*® to expand each of the four components, converted to polar form, in a Taylor series (see (2.16)). The result shows that

$$P[r] = \begin{pmatrix} O[1/r] & O[1/r] \\ O[1/r] & O[1/r] \end{pmatrix}.$$

The comments made in Remark 4.5 also apply to the matrix P.

4.8 Example. By taking once more $(x1, x2) = (0.5, 0.5)$, we can illustrate graphically how each element of $P[\{0.5, 0.5\}, \{y1, y2\}]$ behaves when the singular point is in S. The graphs in Fig. 4.3 reveal the existence of a strong singularity at $(y1, y2) = (x1, x2) = (0.5, 0.5)$. Such singularities are especially troublesome in numerical integration, but, fortunately, this is not an issue in our case when x is off the boundary.

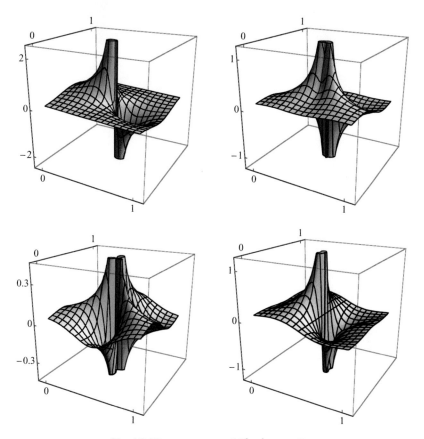

Fig. 4.3 The components of $P[x, y]$ $x, y \in S$.

The presence of a strong singularity on the boundary, however, gives rise to difficulties.

4.9 Example. Suppose that a segment of the line $y1 = y2$ which contains the point $(x1, x2) = (0.5, 0.5)$ is part of the boundary. The graphs in Fig. 4.4 show the behavior of the four components of $P[x, y]$ in the neighborhood of the point $(y1, y2) = (0.5, 0.5)$. These 3-D graphs are displayed to the left and above the boundary curve and truncated on the opposite side. In this case, the unit normal vector is $v = \left(v1(y), v2(y)\right) = \left(1/\sqrt{2}, -1/\sqrt{2}\right)$.

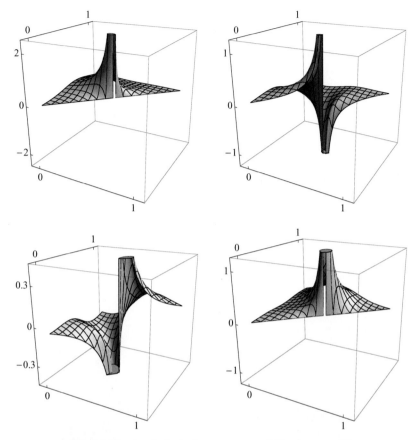

Fig. 4.4 The singularity in the components of $P[x,y]$, x, $y \in \partial S$.

Direct computation indicates that the behavior of the components $P_{\alpha,\beta}[x,y]$, converted to polar coordinates, near $x = y$ on the boundary line $y1 = y2$ is given by

$$P[r] = \begin{pmatrix} O[r] & O[1/r] \\ O[1/r] & O[r] \end{pmatrix}.$$

4.10 Remark. Although $P_{1,1}[x,y]$ and $P_{2,2}[x,y]$ contain factors that suggest the presence of a strong singularity at $x = y$ on ∂S, analytically this is not the case, since the combinations of these factors with the components of the normal derivative turn the leading diagonal elements of $P[x,y]$ into bounded functions when both x and y are on the boundary. In fact, as already mentioned above, these functions tend to 0 at $x = y$ with order $O[r]$. However, numerical techniques never calculate $P[x,y]$ exactly on ∂S (see Remark 4.5), and the graphs of $P_{1,1}[x,y]$ and $P_{2,2}[x,y]$ shown in Fig. 4.4 indicate that if y is slightly off the boundary in the vicinity of $(0.5, 0.5)$, it is close to a strong singularity. This, along with the strongly singular behavior of the other two components, creates significant problems for numerical integration. The difficulty is made worse by the fact that the internal definitions of D and P are much more complicated than what their graphs suggest. These issues are discussed in much more detail below.

4.5 The Singularities of D and P

4.11 Example. We re-examine the behavior of $D[x,y]$ near its singularity by considering the earlier setup, where the singular point is $(x1,x2) = (0.5,0.5)$ and the boundary curve includes a segment of the line $y1 = y2$ containing this point. The values assigned for this illustration are

$$\lambda \to 1, \quad \mu \to 2, \quad k \to 3.$$

Figure 4.5 shows the graphs of the $D_{\alpha,\beta}[x,y] = D_{\alpha,\beta}[\{0.5,\,0.5\},\{y1,\,y2\}]$ along the boundary segment

$$y1 = y2, \quad 0 \le y1 \le 1.$$

We notice the logarithmic singularity in $D_{1,1}[x,y]$ and $D_{2,2}[x,y]$ and the fact that the other two components do not have a jump discontinuity at $(y1, y2) = (0.5, 0.5)$.

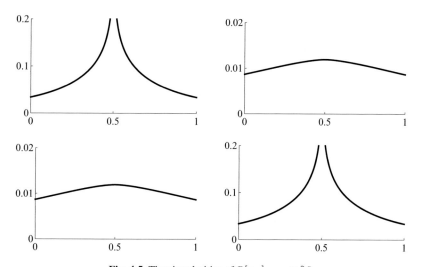

Fig. 4.5 The singularities of $D[x,y]$, $x, y \in \partial S$.

Since the algebraic structure of D is very complicated, it is desirable to make a number of simplifying assumptions so that a more detailed examination becomes possible near the singularity. Thus, we consider the special case where the two-dimensional domain of D is rendered one-dimensional by considering $D[x,y]$ to be defined on the boundary segment $y1 = y2$ with the singularity at the point $(x1, x2) = (0.5, 0.5)$ on it; that is, we make the replacements

$$(x1, x2) \rightarrow (0.5, 0.5),$$

$$y2 \rightarrow y1,$$

$$\sqrt{(x1 - y1)^2 + (x2 - y2)^2} \rightarrow r.$$

This changes $D[x,y]$ to $D[r]$, and we can now perform a single-variable analysis on the new version. First, consider the dominant terms in the Taylor series expansion of $D[r]$:

$$\begin{pmatrix} \dfrac{-7\text{EulerGamma} - 2\text{Log}\left[\frac{\sqrt{\frac{3}{5}}}{2}\right] - 5\text{Log}\left[\frac{\sqrt{\frac{3}{2}}}{2}\right] - 7\text{Log}[r]}{40\pi} + O[r]^1 & \dfrac{3}{80\pi} + O[r]^1 \\[3em] \dfrac{3}{80\pi} + O[r]^1 & \dfrac{-7\text{EulerGamma} - 2\text{Log}\left[\frac{\sqrt{\frac{3}{5}}}{2}\right] - 5\text{Log}\left[\frac{\sqrt{\frac{3}{2}}}{2}\right] - 7\text{Log}[r]}{40\pi} + O[r]^1 \end{pmatrix}.$$

For small values of r, we have the asymptotic behavior (4.7), which confirms that the leading diagonal terms have a logarithmic singularity and that the other two are bounded as $r \rightarrow 0$.

It is important to bear in mind that the above results have been calculated only for a specific case; hence, their simplicity disguises the underlying difficulties encountered in the construction of D. For example, $D[r]$ contains modified Bessel functions, which have a singularity at $r = 0$:

$$\text{BesselK}[0, r] \rightarrow \text{Log}[r],$$

$$\text{BesselK}[1, r] \rightarrow 1/r,$$

$$\text{BesselK}[2, r] \rightarrow 1/r^2.$$

These functions are combined with other algebraic expressions in such a way that all but the logarithmic singularities cancel out in the leading diagonal entries as $r \rightarrow 0$, and all the singularities in the other two components are eliminated. However, when D is evaluated numerically, the expressions involving D must be evaluated individual part by individual part before they are amalgamated into a final result. A detailed examination of $D_{\alpha,\beta}$ indicates that, for small values of r, the internal structure of D has the asymptotic behavior

$$D_{\alpha,\beta}[r] = O[\text{Log}[r]] + \frac{O[r]O[1/r]}{O[r^2]} + \frac{O[r^2]O[1/r]}{O[r^3]} + \frac{O[r^2]O[1/r^2]}{O[r^2]} + O[1/r^2].$$

$$(4.11)$$

4.12 Remark. Formula (4.11) applies to the specific algebraic representation used in our code and may change if another such representation is chosen.

This shows that, individually, the various parts of D have a much stronger singular behavior than what we expect when they are assembled together after internal cancelation, which makes the numerical calculation considerably more inaccurate than would otherwise be the case. It is a well-understood concept in numerical analysis that the difference of nearly equal large expressions generates a significant loss of accuracy when floating point arithmetic is used. The problem is exacerbated by the higher-than-expected order of the singularities in the individual expressions being evaluated. The worst possible asymptotic scenario when these components are calculated before internal cancelation takes place is

$$D[r] = \begin{pmatrix} O[1/r^3] & O[1/r^3] \\ O[1/r^3] & O[1/r^3] \end{pmatrix}.$$

We can convince ourselves of the loss of computational accuracy in the neighborhood of $r = 0$ if we divide the components of $D[r]$ by the first term in their corresponding Taylor series. Then we should have

$$\text{Limit}\left[\begin{pmatrix} \dfrac{D_{1,1}[r]}{\text{first term}} & \dfrac{D_{1,2}[r]}{\text{first term}} \\ \dfrac{D_{2,1}[r]}{\text{first term}} & \dfrac{D_{2,2}[r]}{\text{first term}} \end{pmatrix}, r \to 0 \right] = \begin{pmatrix} 1 & 1 \\ 1 & 1 \end{pmatrix}.$$

Graphing the four components for $0.0 \le r \le 0.00001$ shows the numerical instability of the computation near $r = 0$ (see Fig. 4.6).

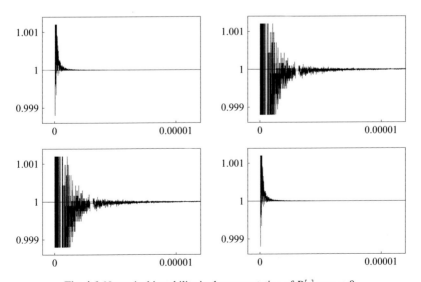

Fig. 4.6 Numerical instability in the computation of $D[r]$ as $r \to 0$.

The graphs in Fig. 4.6 indicate a significant loss of accuracy for small values of
r. In this case, the loss amounts to more than 10 digits of relative accuracy, which
has a severe adverse effect on the numerical evaluation of integrals. In the boundary element method, we expect the largest contribution to the value of our definite
integrals to occur in the vicinity of the singular point. Unfortunately, this is exactly
where the greatest numerical inaccuracy also occurs.

At a first glance, a relative error of about 0.1% should be easy to handle with
a small increase in computational working precision. But this is not necessarily
the case. The matrix D is used in the boundary element method as the kernel of a
definite integral along the boundary (see (4.8)). Typically, numerical quadrature is
of the form

$$\sum_i w_i f(x_i),$$

where the w_i are the weights corresponding to the quadrature method and the x_i are
the quadrature points. In cases involving D, the numbers $f(x_i)$ become extremely
large for x_i near the singularity. Therefore, although the relative error in the $f(x_i)$ is
small, the values of $f(x_i)$ when the x_i are in the neighborhood of the singular point
can significantly contaminate the result of the numerical quadrature.

Numerical integration issues raised by singularities are discussed at length in
Sect. 4.6.

4.13 Remark. Numerical inaccuracy can be partially mitigated by

   (i)   increasing the internal working precision of *Mathematica*®;
  (ii)   simplifying the expression of D to force as much internal cancelation as pos-
         sible;
 (iii)   eliminating the higher-order modified Bessel functions, which contain the
         higher-order singularities.

The default floating point working precision is machine-dependent, with a common value of 16 digits. The graphs in Fig. 4.6 have been generated with this value.
If the working precision is increased by 10 digits—that is, from 16 to 26—then
most of the numerical computational inaccuracy disappears. The graphs for $D[r]$ in
Fig. 4.7, produced by the same algorithm but with this increased working precision,
confirm the expected improvement.

Simplification in *Mathematica*® can be performed by means of the functions
Simplify and FullSimplify. They both operate by applying appropriate
transformation rules to their argument and returning the simplest possible expression for it. One problem here is that the number of possible simplification paths that
must be explored grows tremendously fast with the size of the input expression. As
a consequence, the process can take an excessive amount of time to reach comple-
tion. When large expressions are encountered, such as one of the components of
D, it is usually necessary to segment the problem into smaller pieces and simplify
these preliminary results before simplifying the full expression. Simplify and
FullSimplify often need additional information to complete the process. This
is supplied by inserting suitable assumptions in the program.

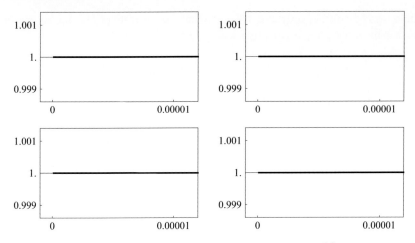

Fig. 4.7 Improved numerical stability near the singularities of $D[r]$ as $r \to 0$.

4.14 Example. The simple algebraic expression $\sqrt{r^2}$ that appears repeatedly within $D[r]$ simplifies differently depending on the assumption on the nature of r. Specifically,

$\sqrt{r^2}$ simplifies to r if we assume that $r \in$ Reals and $r > 0$;

$\sqrt{r^2}$ simplifies to $\text{Abs}[r]$ if we assume that $r \in$ Reals;

$\sqrt{r^2}$ simplifies to $\sqrt{r^2}$ if we assume that $r \in$ Complex, which is the default.

4.15 Remark. The function `Simplify` is restricted to the most commonly used simplification procedures, usually consisting of algebraic, exponential, and trigonometric transformation rules. The expressions of D and P contain additional functions, including the modified `BesselK` functions. Identities related to the `BesselK` are not included in the structure of `Simplify`, so this function will not achieve the desired simplification. In such cases we need to use the function `FullSimplify`, which contains a much larger set of transformations but is correspondingly much slower in its evaluation.

4.16 Example. The function `Simplify` is inadequate for

$r\text{BesselK}[0,r] - 2\text{BesselK}[1,r] + r\text{BesselK}[2,r],$

whereas `FullSimplify` produces the result

$2r\text{BesselK}[0,r].$

The original expression of D is very large, with a `LeafCount` of 2,943. However, after extensive simplification and forcing the elimination of the higher-order modified Bessel function $\text{BesselK}[2,r]$, it can be reduced to a much smaller one, with a LeafCount of only 782. Also, the internal structure of D at the singularity now has the more computationally friendly form

$$D[r] = \begin{pmatrix} O[\mathrm{Log}[r]] + \dfrac{O[r]O[r^{-1}]}{O[r^2]} & O[\mathrm{Log}[r]] + O[r]O[r^{-1}] \\[2ex] O[\mathrm{Log}[r]] + O[r]O[r^{-1}] & O[\mathrm{Log}[r]] + \dfrac{O[r]O[r^{-1}]}{O[r^2]} \end{pmatrix}.$$

As mentioned in Remark 4.12, this expression corresponds to the algebraic form chosen for our code and, as such, is not unique.

Next, we consider the algebraic structure of P, which is even more complicated than that of D and was examined in the previous section.

4.17 Example. We go back to the earlier special case (see Example 4.11), where the singularity is at $(x1, x2) = (0.5, 0.5)$ and the boundary curve includes a segment of the line $y1 = y2$ that contains the singular point. The other parameter values are unchanged, and the unit normal vector is $(v1, v2) = (1/\sqrt{2}, -1/\sqrt{2})$.

The graphs in Fig. 4.8 illustrate the behavior of the components

$$P_{\alpha,\beta}[x, y] = P_{\alpha,\beta}[\{0.5, 0.5\}, \{y1, y2\}]$$

along the boundary segment

$$y1 = y2, \quad 0 \le y1 \le 1.$$

We notice the strong singularity of $P_{1,2}[x, y]$ and $P_{2,1}[x, y]$, and the absence along the boundary path of any singularity or discontinuity at $(y1, y2) = (0.5, 0.5)$ in the other two components.

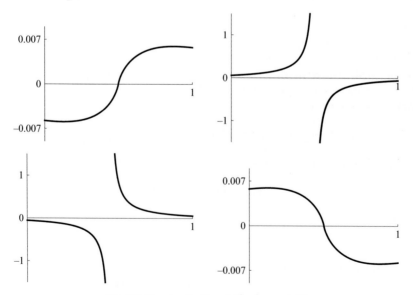

Fig. 4.8 The singularities of $P[x, y]$, $x, y \in \partial S$.

The analysis of P is in many respects similar to that of D, but there are also some important differences between the two. After all the internal cancelations, the theoretical order of the singularities in P when y lies on the boundary is

$$P[r] = \begin{pmatrix} O[r] & O[r^{-1}] \\ O[r^{-1}] & O[r] \end{pmatrix}.$$

Close examination of P reveals that it contains the modified Bessel function $\mathrm{BesselK}[3, r]$, which is of order r^{-3} for small values of r. As a consequence, P is constructed from parts with worse singularities than those in D. A thorough check of their orders before internal cancelation indicates that, as $r \to 0$,

$$P_{1,1}[r] = \frac{O[r]O[\mathrm{Log}[r]]}{O[r^2]} + \frac{O[r^3]O[\mathrm{Log}[r]]}{O[r^4]} + \frac{O[r]O[r^{-1}]}{O[r]} + \frac{O[r]O[r^{-1}]}{O[r^3]}$$
$$+ \frac{O[r^3]O[r^{-1}]}{O[r^3]} + \frac{O[r]O[r^{-2}]}{O[r^2]} + \frac{O[r^3]O[r^{-2}]}{O[r^4]} + \frac{O[r^3]O[r^{-3}]}{O[r^3]},$$

$$P_{1,2}[r] = \frac{O[r]O[\mathrm{Log}[r]]}{O[r^2]} + \frac{O[r^3]O[\mathrm{Log}[r]]}{O[r^4]} + \frac{O[r]O[r^{-1}]}{O[r]} + \frac{O[r^3]O[r^{-1}]}{O[r^3]}$$
$$+ \frac{O[r]O[r^{-2}]}{O[r^2]} + \frac{O[r^3]O[r^{-2}]}{O[r^4]} + \frac{O[r^3]O[r^{-3}]}{O[r^3]},$$

$$P_{2,1}[r] = \frac{O[r]O[\mathrm{Log}[r]]}{O[r^2]} + \frac{O[r^3]O[\mathrm{Log}[r]]}{O[r^4]} + \frac{O[r]O[r^{-1}]}{O[r]} + \frac{O[r^3]O[r^{-1}]}{O[r^3]}$$
$$+ \frac{O[r]O[r^{-2}]}{O[r^2]} + \frac{O[r^3]O[r^{-2}]}{O[r^4]} + \frac{O[r^3]O[r^{-3}]}{O[r^3]},$$

$$P_{2,2}[r] = \frac{O[r]O[\mathrm{Log}[r]]}{O[r^2]} + \frac{O[r^3]O[\mathrm{Log}[r]]}{O[r^4]} + \frac{O[r]O[r^{-1}]}{O[r]} + \frac{O[r]O[r^{-1}]}{O[r^3]}$$
$$+ \frac{O[r^3]O[r^{-1}]}{O[r^3]} + \frac{O[r]O[r^{-2}]}{O[r^2]} + \frac{O[r^3]O[r^{-2}]}{O[r^4]} + \frac{O[r^3]O[r^{-3}]}{O[r^3]}.$$

The statement made in Remark 4.12 applies here as well.

As in the case of D, we see that the individual parts of P have a much stronger singular behavior than that of the final result obtained by combining them. The worst possible asymptotic scenario here for the numerical singular behavior of $P[r]$ is

$$\begin{pmatrix} O\left[r^{-4}\right] & O\left[r^{-4}\right] \\ O\left[r^{-4}\right] & O\left[r^{-4}\right] \end{pmatrix}.$$

To show graphically the loss of computational accuracy for $P[r]$ as $r \to 0$, we note that $P_{1,1}[r]$ and $P_{2,2}[r]$ have a zero of order $O[r]$ and are, therefore, bounded, whereas the other two components, which exhibit singular behavior, are unbounded. The latter can be divided by the singular portion of the first term in their Taylor series, namely

$$(1/(5\sqrt{2}\pi))r^{-1} = O[r^{-1}] \text{ for } P_{1,2}[r],$$
$$-(1/(5\sqrt{2}\pi))r^{-1} = O[r^{-1}] \text{ for } P_{2,1}[r],$$

to generate bounded results. The expected outcome for all the components of P is that they should tend to either 0 or 1 as $r \to 0$; that is,

$$\text{Limit}\left[\left(\begin{array}{cc} P_{1,1}[r] & \dfrac{P_{1,2}[r]}{\text{first term}} \\ \dfrac{P_{2,1}[r]}{\text{first term}} & P_{2,2}[r] \end{array}\right), r \to 0\right] = \begin{pmatrix} 0 & 1 \\ 1 & 0 \end{pmatrix}.$$

Figure 4.9 shows the numerical instability of the computation of the four components for small values of r.

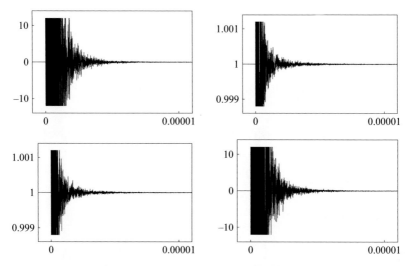

Fig. 4.9 Numerical instability for $P[r]$ as $r \to 0$.

The graphs in Fig. 4.9, generated for the same range of r that we used for $D[r]$, indicate a much greater loss of accuracy than in the case of $D[r]$. This is especially relevant for the comparison with $D_{1,1}[r]$ and $D_{2,2}[r]$, which have a magnitude of at most 0.006 but produce errors with magnitudes significantly larger than 12. Such an extreme relative error may seem surprising at first, until we realize that $D_{1,1}[0]$ and $D_{2,2}[0]$ are close to an off-boundary singularity in the interior of the domain, with theoretical order $O[r^{-1}]$ and numerical order $O[r^{-4}]$. It is therefore obvious that we need to develop special strategies for handling numerical difficulties of this nature successfully.

The procedures for reducing the numerical inaccuracy mentioned in Remark 4.13 apply in equal measure to P. For example, Fig. 4.9 has been generated with 16-digit working precision. Increasing the working precision by 10 digits causes most of the computational inaccuracy to disappear, as can be seen from Fig. 4.10, which has been produced with the same algorithm but with 26-digit working precision.

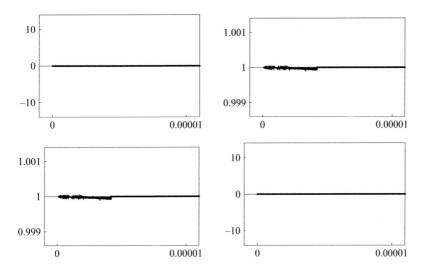

Fig. 4.10 Improved numerical stability for $P[r]$ as $r \to 0$.

The matrix P can be simplified considerably. After that, and after the functions BesselK$[2, r]$ and BesselK$[3, r]$ are forcibly removed, we can reduce the LeafCount for P from 29,183 to 3,477. This shrinks the size of P by about a factor of ten. Also, the internal structure of the singularity in P now has the more computationally friendly asymptotic form

$$P_{1,1}[r] = \frac{O[r]O[\text{Log}[r]]}{O[r^2]} + \frac{O[r^3]O[\text{Log}[r]]}{O[r^4]} + \frac{O[r]O[r^{-1}]}{O[r]} + \frac{O[r]O[r^{-1}]}{O[r^3]}$$
$$+ \frac{O[r^3]O[r^{-1}]}{O[r^3]},$$

$$P_{1,2}[r] = \frac{O[r]O[\text{Log}[r]]}{O[r^2]} + \frac{O[r^3]O[\text{Log}[r]]}{O[r^4]} + \frac{O[r]O[r^{-1}]}{O[r]} + \frac{O[r^3]O[r^{-1}]}{O[r^3]},$$

$$P_{2,1}[r] = \frac{O[r]O[\text{Log}[r]]}{O[r^2]} + \frac{O[r^3]O[\text{Log}[r]]}{O[r^4]} + \frac{O[r]O[r^{-1}]}{O[r]} + \frac{O[r^3]O[r^{-1}]}{O[r^3]},$$

$$P_{2,2}[r] = \frac{O[r]O[\text{Log}[r]]}{O[r^2]} + \frac{O[r^3]O[\text{Log}[r]]}{O[r^4]} + \frac{O[r]O[r^{-1}]}{O[r]} + \frac{O[r]O[r^{-1}]}{O[r^3]}$$
$$+ \frac{O[r^3]O[r^{-1}]}{O[r^3]}.$$

However, the order of the computational singularity is still higher than the expected theoretical order of $O[r^{-1}]$.

This analysis is used later in the chapter to determine how to handle numerical integration issues that arise in the boundary element method. These issues include

(i)   the ability to handle logarithmic singularities;
(ii) the ability to handle Cauchy principal value integration;
(iii) the ability to increase computational accuracy to compensate for the inaccuracy introduced by the behavior of D and P near their singularities.

4.6 Numerical Integration

Anyone who spends a lot of time working with singular integrals eventually spends a lot of time struggling with numerical integration techniques. In general, the algorithms coded in the `NIntegrate` function are near optimal for generic purposes, and any attempt to create one's own algorithms usually results in disappointment. However, we have to do exactly that since in solving our problems we need to handle both weak and strong singularities.

The first consideration is the degree of desired accuracy in the computed result. This is ensured by specifying the *Mathematica*® parameters `AccuracyGoal`, `PrecisionGoal`, and `WorkingPrecision`. At a first glance, it might appear that using machine precision (as mentioned earlier, normally about 16 digits) would be sufficient to meet our needs. But the definite integrals of functions with singularities present special problems. Below, we show in greater detail that there is often a significant loss of accuracy in the computation of integrals of such functions.

4.18 Example. The standard strategy for computing the Cauchy principal value of the integral over an interval (a,b) of a function f with a strong singularity at a point c, $a < c < b$, is to use the decomposition

$$\int_a^b f(t)\,dt = \int_a^{c-\varepsilon} f(t)\,dt + \int_0^\varepsilon (f(c+\xi)+f(c-\xi))\,d\xi + \int_{c+\varepsilon}^b f(t)\,dt, \qquad (4.12)$$

dictated by the necessity to integrate f on both sides of the singularity over a symmetric interval of length ε. The two functions $f(c+\xi)$ and $f(c-\xi)$ have nearly equal extremely large absolute values but are of opposite signs on $0 < \xi < \varepsilon$, which leads to a significant loss in precision since most of the significant digits for each function cancel out during the subtraction. The remaining digits are of much lower accuracy. The closer we are to the singularity, the more significant digits we lose. To compensate for this effect, an increase in computational accuracy along with helpful transformations must be deliberately built into the method.

The parameters that control accuracy and precision are

$$\text{PrecisionGoal} \to p, \quad \text{AccuracyGoal} \to a,$$

and they tell *Mathematica*® to attempt to make the numerical error in a result x to be less than $10^{-a} + |x| 10^{-p}$. This would appear to be very straightforward, except that special problems occur in certain circumstances.

4.19 Example. Many of the spline basis functions used to approximate the solution are numerically zero on certain subintervals of the boundary. In these cases, it is clear that the `PrecisionGoal` for the piece of the integral over that specific subinterval can never be met because x is 0. A lot of computational time can be consumed by numerically integrating

$$\int_{t_i}^{t_{i+1}} 0.0 \, dt,$$

which produces the error message "Integral and error estimates are 0 on all integration subregions. Try increasing the value of the `MinRecursion` option. If value of integral may be 0, specify a finite value for the `AccuracyGoal` option."

Using a value for `AccuracyGoal` would appear to be a good solution, but it may be unclear ahead of time exactly what an acceptable magnitude of the definite integrals being evaluated might be. This can cause us to set an unacceptable error tolerance, either much too large or unrealistically small. The best way to handle this is to preprocess the definite integral symbolically and set its value to 0 on the relevant subinterval before it is sent to the numerical integration algorithm.

Ill-conditioning is another issue encountered when using numerical methods. Frequently, the default machine accuracy of 16 digits is insufficient to overcome it, and we are forced to increase the `WorkingPrecision` option to get a satisfactory resolution. This, however, presents a new problem. As all *Mathematica*® programmers know, a floating point expression will automatically assume the precision of the least precise subexpression it contains. Consequently, entering, say, $1/2$ as 0.5 eliminates any possibility of computing the result with greater accuracy than is available with the default machine precision. Since it may be unclear at the beginning how much additional precision is required to achieve our objectives, it is best for us to start with exact arithmetic for all subexpressions and then, if necessary, increase the working precision beyond the default machine precision when the computationally sensitive portion of the program is reached. Once the computationally intensive part has been completed, we can adjust the working precision downward and thus save computational time. Exact arithmetic is easy to code. For example, the machine accurate floating-point number 3.1415 would be entered as the infinite precision rational number $31415/10000$.

Another issue is overcoming the loss of accuracy in computing integrals in the vicinity of a singular point. This usually requires increasing the value assigned to the `WorkingPrecision` option even more. Normally, an increased working precision is required only for the evaluation of the integral over the subinterval containing the singularity; maintaining it outside that region would significantly slow down the computation of all the other results. We take the view that determining

the necessary additional increase in working precision is a matter of experience, as
well as trial and error. Often, an additional 30 digits of working precision accuracy
are required to achieve the `PrecisionGoal` specified. This could result in a total
working precision exceeding 40 digits. The code used in the creation of the material
for this book has been made adaptive in the sense that if the `PrecisionGoal` has
not been met, then the current working precision is automatically increased, and the
result is recomputed until convergence occurs with the desired accuracy.

In summary, our objective is to take the interval (a,b) and divide it into subintervals using the procedure described above, so that $f(t)$ is analytic in each of the
subintervals (a_j,b_j); that is,

$$\int_a^b f(t)\,dt = \sum_{j=1}^{noIntervals} \int_{a_j}^{b_j} f(t)\,dt.$$

Next, the function `NIntegrate` numerically evaluates the corresponding sub-
integrals by means of adaptive techniques, until the preset precision goal is met for
each of them; specifically,

$$\int_a^b f(t)\,dt = \sum_{j=1}^{noIntervals} \int_{a_j}^{b_j} f(t)\,dt \cong \sum_{j=1}^{noIntervals} \sum_{i=1}^{order} w_{i,j} f(t_{i,j}).$$

Special attention must be paid to the subinterval containing the singularity, which
needs to be handled on its own. Consider, for example, a Cauchy principal value
integral with the singularity in the integrand f at $t = c$, $a_j < c < b_j$. By (4.12),

$$\int_{a_j}^{b_j} f(t)\,dt = \int_{a_j}^{c-\varepsilon} f(t)\,dt + \int_0^\varepsilon (f(c+\xi)+f(c-\xi))\,d\xi + \int_{c+\varepsilon}^{b_j} f(t)\,dt.$$

The value of ε is chosen to be a fraction of the distance from the singular point
to the nearest point on the `Exclusions` list compiled for `NIntegrate`, which
comprises all the singularities, the corners that the boundary might possess, and the
knots used in any spline approximation.

The two integrals

$$\int_{a_j}^{c-\varepsilon} f(t)\,dt, \quad \int_{c+\varepsilon}^{b_j} f(t)\,dt$$

are evaluated by `NIntegrate` in the usual way. The singular integral

$$\int_0^\varepsilon (f(c+\xi)+f(c-\xi))\,d\xi$$

is now evaluated separately by `NIntegrate` using increased working precision
combined with singularity-handling adaptive numerical quadrature techniques. In

our experience, the double exponential quadrature method yields the fastest and most accurate result for the type of singularities encountered in the boundary element technique. This method transforms the original integrand in a way that suppresses the singularity and subsequently evaluates the integral numerically using the trapezoidal rule. The transformation used is specifically chosen because its derivative decreases doubly exponentially when the variable of the integrand reaches 0 and ε. In our case, a double exponential transformation of the form $\xi = \phi(\eta)$ changes the integral

$$\int_0^\varepsilon g(\xi)\,d\xi,$$

where

$$g(\xi) = f(c+\xi) + f(c-\xi),$$

to

$$\int_{-\infty}^{+\infty} g(\phi(\eta))\phi'(\eta)\,d\eta. \tag{4.13}$$

The integrand $g(\xi)$ must be analytic in $(0,\varepsilon)$ and may have a singularity at 0. The analyticity condition is satisfied because of the way we have constructed the `Exclusions` for `NIntegrate`. The singularity in $g(\xi)$ must be weaker than $O(\xi^{-1})$ for the integral to converge. Here, the function $g(\xi) = f(c+\xi) + f(c-\xi)$, which cancels the strong $O(\xi^{-1})$ singularity, still has a logarithmic singularity of order $O(\text{Log}[\xi])$ as $\xi \to 0$. The logarithmic singularity can be compensated for by the use of the double exponential quadrature method because the transformed integrand decreases doubly exponentially; that is,

$$g(\phi(\eta))\phi'(\eta) \approx \exp(-c\exp(|\eta|)) \quad \text{as } \eta \to \pm\infty.$$

The function $\phi(\eta)$ is analytic in $(-\infty,\infty)$. The trapezoidal rule was chosen because it is known to be optimal for this type of situation. The transformation used in our case, where the singularity is located at $\xi = 0$, is

$$\xi = \tfrac{1}{2}\varepsilon + \tfrac{1}{2}\varepsilon \tanh\left(\tfrac{1}{2}\pi\sinh\eta\right) \tag{4.14}$$

and converts $\xi = 0$ into $\eta \to -\infty$ and $\xi = \varepsilon$ into $\eta \to \infty$. The trapezoidal rule applied to (4.13) yields

$$\int_{-\infty}^{\infty} g(\phi(\eta))\phi'(\eta)\,d\eta \cong h \sum_{i=-\infty}^{\infty} f(\phi(ih))\phi'(ih).$$

The terms in the trapezoidal rule decay doubly exponentially as $i \to \pm\infty$; therefore, the infinite sum in it is truncated when the computed terms become too small to make a significant contribution to the total.

4.20 Example. The construction of $P[x,y[t]]$ usually guarantees that a logarithmic singularity still remains even after the cancelation of the strong singularity in the Cauchy principal value integration method. Suppose that we want to evaluate

$$\int_0^1 \text{Log}[\xi]\,d\xi.$$

The graph of the integrand $\text{Log}[\xi]$ for $0 < \xi \le 1$ is on the left in Fig. 4.11, and the graph of the function (4.14) for $-\infty < \eta < \infty$ is on the right.

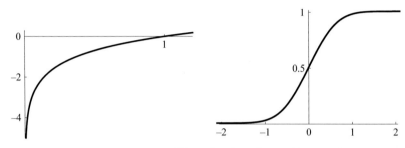

Fig. 4.11 Graphs of $\text{Log}[\xi]$ and $\xi = \frac{1}{2}\varepsilon + \frac{1}{2}\varepsilon\tanh\left(\frac{1}{2}\pi\sinh\eta\right)$.

The transformed integral is

$$\int_{-\infty}^{\infty} g(\phi(\eta))\phi'(\eta)\,d\eta = \int_{-\infty}^{\infty} \text{Log}(\phi(\eta))\phi'(\eta)\,d\eta.$$

The function $\phi'(\eta)$ for $\varepsilon = 1$, which contains the double exponential decay, is

$$\phi'(\eta) = \frac{1}{4}\pi\text{Cosh}[\eta]\text{Sech}\left[\frac{1}{2}\pi\text{Sinh}[\eta]\right]^2,$$

graphed on the left in Fig. 4.12. We remark that this function decays very quickly, as seen from the values

$$\phi'(\pm 3) = 6.8 \times 10^{-13}, \quad \phi'(\pm 4) = 5.0 \times 10^{-36}.$$

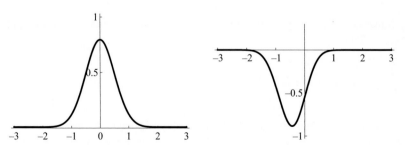

Fig. 4.12 The double exponential decay transformation.

The graph on the right in Fig. 4.12 represents the integrand

$$g(\phi(\eta))\phi'(\eta)$$
$$= \left(\text{Log}\left[\tfrac{1}{2} + \tfrac{1}{2}\text{Tanh}\left[\tfrac{1}{2}\pi\text{Sinh}[\eta]\right]\right]\right)\left(\tfrac{1}{4}\pi\text{Cosh}[\eta]\text{Sech}\left[\tfrac{1}{2}\pi\text{Sinh}[\eta]\right]^2\right).$$

We notice that

$$g(\phi(3))\phi'(3) = -2_1 \times 10^{-14} * 6.8 \times 10^{-13},$$

and that the logarithmic singularity has been completely overpowered.

Example 4.20 illustrates three important features of the double exponential quadrature method:

(i)   The domain of numerical integration is finite, not infinite, and is usually quite narrow.

(ii) Additional digits of accuracy can be achieved with a very small increase in the size of the domain of numerical integration.

(iii) The singularity in the integrand has been removed.

The first feature implies that the number of functional evaluations required in the trapezoidal rule can be reduced, which increases the computational speed of the method. The second one means that adaptive quadrature requires very little additional work to increase its level of accuracy. Finally, the third feature points to the fact that, the singularity having been removed, the trapezoidal rule will not involve operations with numbers of significantly different magnitudes. Combining large and small numbers in floating-point arithmetic usually results in a slowing of the rate of convergence as well as a deterioration in accuracy.

The double exponential quadrature method can be much faster than other quadrature techniques.

4.21 Example. The regular (untransformed) trapezoidal rule has a great deal of difficulty with the singularity in an integral such as

$$\int_0^1 \text{Log}[\xi]\,d\xi. \tag{4.15}$$

We can compare the two methods on the basis of the number of digits of accuracy required in the answer. The bottom set of points on the left in Fig. 4.13 shows that to achieve 1 to 5 digits of relative accuracy, the double exponential quadrature method needs only about 12 to 30 evaluations of $\text{Log}[\xi]$. The top set of points shows that the untransformed trapezoidal rule requires over 1,000,000 evaluations of $\text{Log}[\xi]$ to ensure 5 digits of accuracy. Clearly, the time consumed by the untransformed trapezoidal rule in a case like this is unacceptably excessive.

4.22 Example. We can also compare the double exponential quadrature method to the global adaptive technique, which is the default procedure used by `NIntegrate`.

The latter is a higher-order Gaussian quadrature method with adaptive strategies and singularity-handling capabilities, and it does a reasonable job in evaluating the integral (4.15).

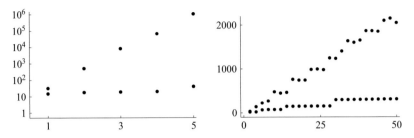

Fig. 4.13 Double exponential quadrature vs. global adaptive method.

The graph on the right in Fig. 4.13 shows the number of evaluations of $\text{Log}[\xi]$ required to achieve 50 digits of relative accuracy for each quadrature method. We see that the double exponential quadrature method (the bottom set of points) does this with about 300 evaluations as opposed to over 2,000 evaluations needed by the global adaptive method (the top set of points).

4.23 Remark. It is useful to know where and how often the integrand $\text{Log}[\xi]$ in (4.15) is evaluated to perform the numerical quadrature. The graph on the left in Fig. 4.14 shows the exact sequence of chosen values of ξ, $0 \le \xi \le 1$. The desired relative accuracy for our answer was 30 digits. We notice that more evaluations occur at the endpoints of the domain, where we anticipate that a singularity might exist.

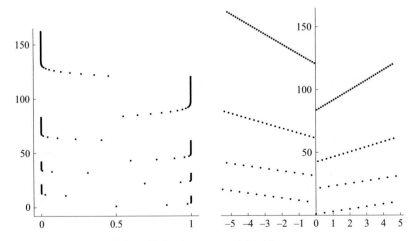

Fig. 4.14 Double exponential quadrature.

The graph on the right shows more clearly the integration scheme in the transformed variable η, $-\infty < \eta < \infty$. First, we see the limited range of η values actually used. Also, we notice the implementation of the trapezoidal rule in the transformed variable. This is recursive, with the number of evaluation points doubled every time the accuracy is insufficient, until convergence occurs.

Next, we investigate how the double exponential quadrature method changes the behavior of D in the vicinity of its singularities. To this end, we use again the form $D[r]$, $0 \leq r \leq 1$, derived for D in Example 4.11. The loss of relative accuracy in the computation of the four components of $D[r]$ can be seen from the graphs for

$$0.0000001 \leq r \leq 0.00001,$$

shown in Fig. 4.15. These graphs indicate numerical instability as $r \to 0$. They were generated with a working precision of 16 digits, and the range of 1 ± 0.001 represents a relative error of $\pm 0.1\%$.

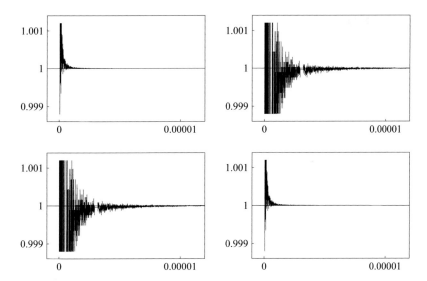

Fig. 4.15 Relative computational error for the untransformed $D[r]$ as $r \to 0$.

The double exponential quadrature method transforms the integral

$$\int_0^1 D(r)\,dr$$

to

$$\int_{-\infty}^{\infty} D(\phi(\eta))\phi'(\eta)\,d\eta,$$

where the integrand, when generated by the double exponential transformation (4.14), is

$$D(\phi(\eta))\phi'(\eta)$$
$$= \left(D\left[\tfrac{1}{2} + \tfrac{1}{2}\mathrm{Tanh}\left[\tfrac{1}{2}\pi\mathrm{Sinh}[\eta]\right]\right]\right)\left(\tfrac{1}{4}\pi\mathrm{Cosh}[\eta]\mathrm{Sech}\left[\tfrac{1}{2}\pi\mathrm{Sinh}[\eta]\right]^2\right).$$

The graphs in Fig. 4.16 show the behavior of the transformed matrix $D(\phi(\eta))\phi'(\eta)$.

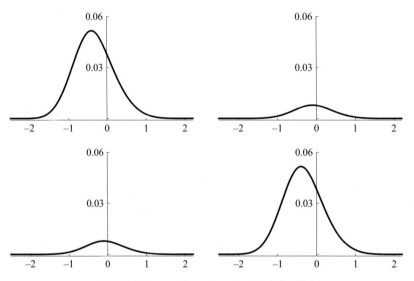

Fig. 4.16 The transformed matrix $D(\phi(\eta))\phi'(\eta)$.

These graphs have been drawn for

$$-2_34 \le \eta \le -2_01,$$

which corresponds to

$$0.0000001 \le r \le 0.00001.$$

We recall that $D[r]$ is not bounded as $r \to 0$, but that its relative size is bounded. The transformed expression $D(\phi(\eta))\phi'(\eta)$ is bounded everywhere.

Dividing each of the four components by their maximum value, we can plot the relative error for the transformed expressions. The graphs in Fig. 4.17 show the relative error in the computation of $D(\phi(\eta))\phi'(\eta)$ over the equivalent domain in η; the range $(0.0, 0.001)$ represents a relative error of $\pm 0.1\%$ of the maximum value for each component. These graphs, generated with a working precision of 16 digits, do not exhibit any of the numerical instability seen earlier in the untransformed $D[r]$ for small values of r.

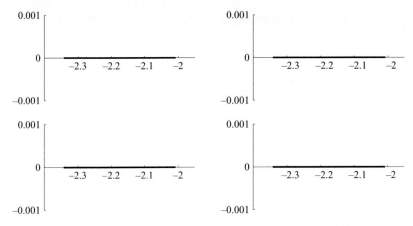

Fig. 4.17 The relative computational error in the transformed $D(\varphi(\eta))\varphi'(\eta)$.

In summary, the double exponential quadrature method represents an ideal choice for performing the numerical integration required by the boundary element method.

4.7 Coding the Single-Layer and Double-Layer Potentials

With the necessary symbolic code now in place, we can formulate the single-layer and double-layer potentials (see Definition 2.3) as

$$(\mathbf{V}\varphi)[x] = \oint_{\Gamma_{\text{Weak}[x]}} D[x,y] \circ \varphi[y] \, d\Gamma_y,$$

$$(\mathbf{W}\psi)[x] = \oint_{\Gamma_{\text{CPV}[x]}} P[x,y] \circ \psi[y] \, d\Gamma_y, \tag{4.16}$$

where $\varphi = \{\varphi_1[y], \varphi_2[y]\}$ and $\psi = \{\psi_1[y], \psi_2[y]\}$ are two-component density functions. When x is on the boundary, $D[x,y]$ and $P[x,y]$ have weak and strong singularities, which the code designed to evaluate the boundary integrals containing these matrices as kernels needs to take into account. When x is not on the boundary, the kernels have no singularities, and normal integration along the boundary can be performed to evaluate the corresponding boundary integrals. With this coding, the Somigliana formula (2.18) with $x \in \partial S$ is written as

$$\oint_{\Gamma_{\text{Weak}[x]}} D[x,y] \circ (\mathrm{T}u)[y] \, d\Gamma_y = \oint_{\Gamma_{\text{CPV}[x]}} P[x,y] \circ u[y] \, d\Gamma_y + \tfrac{1}{2} u[x], \tag{4.17}$$

and with $x \in S$,

$$\oint_{\Gamma} D[x,y] \circ (\mathrm{Tu})[y] \, d\Gamma_y = \oint_{\Gamma} P[x,y] \circ u[y] \, d\Gamma_y + u[x]. \tag{4.18}$$

4.8 Coding the Boundary Integral Methods

After the single-layer and double-layer potentials have been coded, it is straightforward to do the same for each of the boundary integral methods developed in Chap. 3. For brevity, we illustrate only the case of Dirichlet boundary data.

4.24 Example. The direct method for the Dirichlet problem (D$^+$) defined in Sect. 3.2 leads to the boundary integral equation $(\mathscr{D}_D^+)$; that is,

$$V_0(\varphi) = \left(W_0 + \tfrac{1}{2}I\right)\mathscr{P},$$

where $\mathscr{P}$ is the prescribed Dirichlet data function on ∂S. Using (4.17), we code this equation as

$$\oint_{\Gamma_{\mathrm{Weak}[x]}} D[x,y] \circ \varphi[y] \, d\Gamma_y = \oint_{\Gamma_{\mathrm{CPV}[x]}} P[x,y] \circ \mathscr{P}[y] \, d\Gamma_y + \tfrac{1}{2}\mathscr{P}[x], \tag{4.19}$$

where φ is the unknown density representing Tu. Once this density is determined, we can find the solution in S by means of the representation formula (4.18):

$$u[x] = \oint_{\Gamma} D[x,y] \circ \varphi[y] \, d\Gamma_y - \oint_{\Gamma} P[x,y] \circ \mathscr{P}[y] \, d\Gamma_y, \quad x \in S. \tag{4.20}$$

4.25 Example. The classical indirect method for the Dirichlet problem in S, described in Sect. 3.1, leads to the boundary integral equation $(\mathscr{D}_C^+)$; that is,

$$\left(W_0 - \tfrac{1}{2}I\right)\varphi = \mathscr{P},$$

where $\mathscr{P}$ is the prescribed boundary data function on ∂S. This equation is coded in *Mathematica*® as

$$\oint_{\Gamma_{\mathrm{CPV}[x]}} P[x,y] \circ \varphi[y] \, d\Gamma_y - \tfrac{1}{2}\varphi[x] = \mathscr{P}[x],$$

where $\varphi[y]$ is the unknown density. Again, once the density is determined, we can find the solution in S by means of formula (4.16).

The coding shown above omits a lot of detail. Basically, code is developed to convert the boundary integrals into parametric line integrals. The latter are then evaluated numerically, special attention being paid to the boundary singularities of their kernels.

4.9 Outline of the Boundary Element Method

The boundary element method uses certain functions (elements) to approximate the unknown densities of the boundary integral equations, and consists of several steps.

Step 1. We write the symbolic quantities D and P corresponding to the homogeneous problem

$$Z_x \diamond u = 0.$$

Step 2. We construct the single-layer and double-layer potentials

$$(\mathbf{V}\varphi)[x] = \oint_{\Gamma_{\text{Weak}[x]}} D[x,y] \circ \varphi[y] \, d\Gamma_y,$$

$$(\mathbf{W}\psi)[x] = \oint_{\Gamma_{\text{CPV}[x]}} P[x,y] \circ \psi[y] \, d\Gamma_y.$$

Step 3. We code the equation for the specific boundary integral method we want to apply. Thus, equation $(\mathscr{D}_D^+)$ in the direct method for the Dirichlet problem takes the form shown in Example 4.24, namely

$$\oint_{\Gamma_{\text{Weak}[x]}} D[x,y] \circ \varphi[y] \, d\Gamma_y = \oint_{\Gamma_{\text{CPV}[x]}} P[x,y] \circ \mathscr{P}[y] \, d\Gamma_y + \tfrac{1}{2}\mathscr{P}[x],$$

where φ is the unknown density and $\mathscr{P}$ is prescribed on ∂S.

Step 4. We define a parameterization $y[t]$ of the boundary ∂S. This is done by splitting the boundary into segments $(\partial S)_i$ corresponding to the Dirichlet, Neumann, or Robin conditions prescribed on them, respectively, or to any corners the boundary might have. Corners introduce discontinuities in the derivative of the parameterization and, therefore, create potential discontinuities in one or several of $y'[t]$, $v[t]$, $D[x,y[t]]$, and $P[x,y[t]]$.

Step 5. We use the parameterization of the boundary to change the boundary integral equation into a line integral of the form

$$\oint_{\Gamma_{\text{Weak}[x]}} D[x,y[t]] \circ \varphi[y[t]] \mathrm{dsdt}[t] \, dt$$

$$= \oint_{\Gamma_{\text{CPV}[x]}} P[x,y[t]] \circ \mathscr{P}[y[t]] \mathrm{dsdt}[t] \, dt + \tfrac{1}{2}\mathscr{P}[x].$$

Step 6. We construct approximate solutions $\tilde{\varphi}$ by using the parameterization $y[t]$ and choosing the knots $t_{i,j}$ and the spline degree required to make up a B-spline basis $\{b_{i,j}\}$. The location of the knots is dictated by the boundary segments $(\partial S)_i$. Additional knots within each $(\partial S)_i$ are also selected in order to achieve the desired approximating properties. The approximate solutions are represented as linear combinations of the $b_{i,j}$ with unknown coefficients $c_{\alpha,i,j}$; that is,

$$
\tilde{\varphi}[y[t]] = \begin{pmatrix} \sum_i \sum_j c_{1,i,j} b_{i,j}[t] \\ \sum_i \sum_j c_{2,i,j} b_{i,j}[t] \end{pmatrix}.
$$

Step 7. We substitute the spline elements in the boundary integral equation. For example, equation $(\mathscr{D}_D^+)$ in the direct method for the Dirichlet problem becomes

$$
\oint_{\Gamma_{\text{Weak}[x]}} D[x,y[t]] \circ \begin{pmatrix} \sum_i \sum_j c_{1,i,j} b_{i,j}[t] \\ \sum_i \sum_j c_{2,i,j} b_{i,j}[t] \end{pmatrix} \mathrm{dsdt}[t]\, dt
$$

$$
\cong \oint_{\Gamma_{\text{CPV}[x]}} P[x,y[t]] \circ \mathscr{P}[y[t]] \mathrm{dsdt}[t]\, dt + \tfrac{1}{2}\mathscr{P}[x].
$$

This converts our boundary integral method, which is exact, to a boundary element method, which is approximate.

Step 8. Because *Mathematica*® is a list-based language, the bookkeeping does not become an issue. The list containing the B-spline basis $\{b_{i,j}\}$ for each of the two components of $\tilde{\varphi}$ can be combined into a single list twice as long; more precisely,

$$
\dots, \begin{pmatrix} b_{i,j}[t] \\ 0 \end{pmatrix}, \dots, \begin{pmatrix} 0 \\ b_{i,j}[t] \end{pmatrix}, \dots
$$

is converted to

$$
\begin{pmatrix} \dots\, b_{i,j}[t] \, \dots & 0 & 0 & 0 \\ 0 & 0 & 0 & \dots\, b_{i,j}[t] \, \dots \end{pmatrix}.
$$

The 2×1 matrix equation of the boundary element method at the point x can now be written as

$$
\oint_{\Gamma_{\text{Weak}[x]}} D[x,y[t]] \circ \begin{pmatrix} \dots\, b_{i,j}[t] \, \dots & 0 & 0 & 0 \\ 0 & 0 & 0 & \dots\, b_{i,j}[t] \, \dots \end{pmatrix} \circ \begin{pmatrix} \dots \\ c_{1,i,j} \\ \dots \\ c_{2,i,j} \\ \dots \end{pmatrix} \mathrm{dsdt}[t]\, dt
$$

$$
\cong \oint_{\Gamma_{\text{CPV}[x]}} P[x,y[t]] \circ \mathscr{P}[y[t]] \mathrm{dsdt}[t]\, dt + \tfrac{1}{2}\mathscr{P}[x].
$$

Using the linearity of the integral, we bring this equation to the form

$$\left(\oint\limits_{\Gamma_{\text{Weak}[x]}} D[x,y[t]] \circ \begin{pmatrix} \dots & b_{i,j}[t] & \dots & 0 & 0 & 0 \\ 0 & 0 & 0 & \dots & b_{i,j}[t] & \dots \end{pmatrix} \text{dsdt}[t]\, dt \right) \circ \begin{pmatrix} \dots \\ c_{1,i,j} \\ \dots \\ \dots \\ c_{2,i,j} \\ \dots \end{pmatrix}$$

$$\cong \oint\limits_{\Gamma_{\text{CPV}[x]}} P[x,y[t]] \circ \mathscr{P}[y[t]] \text{dsdt}[t]\, dt + \tfrac{1}{2}\mathscr{P}[x].$$

Step 9. We select a method for computing the unknown coefficients $c_{\alpha,i,j}$. Although there are several methods to choose from, here we use only the collocation technique. We start by selecting the collocation points x_k, $k = 1,\dots,m$, on ∂S and map the 2×1 equation of the boundary element method onto these points to form a linear system of $2m$ equations. The x_k must be chosen so as not to include any corners, as well as to ensure that the resulting linear system has a solution. This system is of the form

$$\begin{pmatrix} \vdots \\ \oint\limits_{\Gamma_{\text{Weak}[x_k]}} D[x_k,y[t]] \circ \begin{pmatrix} \dots & b_{i,j}[t] & \dots & 0 & 0 & 0 \\ 0 & 0 & 0 & \dots & b_{i,j}[t] & \dots \end{pmatrix} \text{dsdt}[t]\, dt \\ \vdots \end{pmatrix} \circ \begin{pmatrix} \vdots \\ c_{1,i,j} \\ \vdots \\ c_{2,i,j} \\ \vdots \end{pmatrix}$$

$$= \begin{pmatrix} \vdots \\ \oint\limits_{\Gamma_{\text{CPV}[x_k]}} P[x_k,y[t]] \circ \mathscr{P}[y[t]] \text{dsdt}[t]\, dt + \tfrac{1}{2}\mathscr{P}[x_k] \\ \vdots \end{pmatrix}.$$

The x_k also include the location of the singularities in $D[x,y[t]]$ or $P[x,y[t]]$. These points must be explicitly passed to the numeric integration algorithms in order to ensure accurate evaluation.

Step 10. The individual components of both the coefficient matrix and the right-hand side vector above are sent for numerical integration. The integration algorithms must accurately handle the singularities of the integrand. This process returns a system with numerical coefficients.

Step 11. The system constructed in the preceding step is either 'square' or overdetermined, and is solved for the unknown coefficients $c_{\alpha,i,j}$.

Step 12. The approximate solution for the unknown density is now represented in terms of the spline basis elements as

$$\tilde{\phi}[y[t]] = \begin{pmatrix} \dots b_{i,j}[t] \ \dots & 0 & 0 & 0 \\ 0 & 0 & 0 & \dots b_{i,j}[t] \ \dots \end{pmatrix} \circ \begin{pmatrix} \vdots \\ c_{1,i,j} \\ \vdots \\ c_{2,i,j} \\ \vdots \end{pmatrix} = \begin{pmatrix} \sum_i \sum_j c_{1,i,j} b_{i,j}[t] \\ \sum_i \sum_j c_{2,i,j} b_{i,j}[t] \end{pmatrix}.$$

Step 13. This approximation is used to determine the approximate solution u in S. In the direct method for (D^+), the representation formula (4.18) yields

$$u[x] \cong \left(\oint_\Gamma D[x,y[t]] \circ \begin{pmatrix} \dots b_{i,j}[t] \ \dots & 0 & 0 & 0 \\ 0 & 0 & 0 & \dots b_{i,j}[t] \ \dots \end{pmatrix} \mathrm{dsdt}[t]\, dt \right) \circ \begin{pmatrix} \vdots \\ c_{1,i,j} \\ \vdots \\ c_{2,i,j} \\ \vdots \end{pmatrix}$$

$$- \oint_\Gamma P[x,y[t]] \circ \mathscr{P}[y[t]] \mathrm{dsdt}[t]\, dt.$$

The rest of this chapter provides additional details for steps 4, 6, and 8.

4.10 Parametrization and Segmentation of the Boundary

Before integrals over ∂S can be evaluated, the boundary curve must be parameterized so that the integrals are converted to ordinary definite integrals with respect to a parameter t. This is done by introducing two suitable functions

$$\{y1,y2\} = \{y1[t], y2[t]\}, \quad 0 \le t \le \mathrm{tMax}.$$

Frequently, these functions are defined piecewise, especially for a boundary with corners. After the parametric form for the normal $\{v1[t], v2[t]\}$ and for $\frac{\mathrm{ds}}{\mathrm{dt}}[t]$ are calculated, the integrals over ∂S are ready for conversion. For example, the Somigliana formula (4.17) with x on the boundary becomes

$$\oint_{\Gamma_{\mathrm{Weak}[x]}} D[x,\{y1[t], y2[t]\}] \circ (\mathrm{Tu})[y1[t], y2[t]] * \mathrm{dsdt}[t]\, d\Gamma_t$$

$$= \oint_{\Gamma_{\mathrm{CPV}[x]}} P[x,\{y1[t], y2[t]\}] \circ u[y1[t], y2[t]] * \mathrm{dsdt}[t]\, d\Gamma_t + \tfrac{1}{2} u[x].$$

Numerical evaluation of integrals over ∂S also requires the boundary to be partitioned into segments on which the integrand is either 'well behaved' or has a well-defined singularity. If the problem has mixed boundary conditions, the boundary curve must first be divided into arcs with only one type of (Dirichlet, Neumann, or Robin) condition prescribed on each of them:

$$\partial S = \bigcup_{i=1,..,m} (\partial S)_i.$$

The boundary integrals need to be evaluated numerically on each of the $(\partial S)_i$. Although the algorithms contained in the `NIntegrate` function are adaptive, they are based on the assumption that the integrand is nearly analytic on each segment. Consequently, the $(\partial S)_i$ must be further subdivided into sub-segments determined by the points where the integrand is not 'well behaved'. This happens in our problems when $D[x,y[t]]$ or $P[x,y[t]]$ or their derivatives have a discontinuity for x on the boundary. Both $u[y[t]]$ and $D[x,y[t]]$ are continuous for $y[t]$ on ∂S except at the singularity $y = x$. Additionally, if the domain has corners, the derivatives of these functions have discontinuities at the corner locations because $y'[t]$ is discontinuous there. Kernels such as $(Tu)[y[t]]$ and $P[x,y[t]]$, whose structure includes the normal, have, therefore, a lower degree of smoothness. Hence, corners must be excluded from the process because the integrands are not sufficiently smooth at these points to meet the conditions for effective and fast numerical integration.

4.26 Example. Let S be the upper half of the disc with center at $(1,0)$ and radius 1, shown on the left in Fig. 4.18. Its boundary, which consists of a half circle and a straight line segment, has two corners and can be parameterized by

$$y1[t] = \begin{cases} 2t, & 0 \le t \le 1, \\ 1 + \mathrm{Cos}[\pi(t-1)], & 1 < t \le 2, \end{cases}$$
$$y2[t] = \begin{cases} 0, & 0 \le t \le 1, \\ \mathrm{Sin}[\pi(t-1)], & 1 < t \le 2. \end{cases} \qquad (4.21)$$

This parameterization starts at the origin and proceeds counterclockwise around the boundary. Here, $D[x,y[t]]$ is continuous for all $0 \le t \le 2$, but its derivatives are discontinuous at the corners—that is, for $t = 0, 1, 2$. On the other hand, $P[x,y[t]]$ is discontinuous at these points, which are shown on the right in Fig. 4.18.

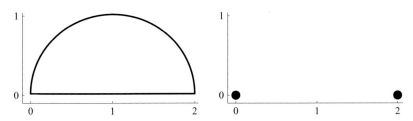

Fig. 4.18 Left: the boundary ∂S. Right: the corners and exclusions for a half disc.

4.27 Remark. The boundary element method requires that either $u[y]$ or $(\mathrm{T}u)[y]$ be replaced by approximating functions $\psi[y]$ or $\varphi[y]$ in the Somigliana formula. For this purpose, our approximating functions are splines, one for each of the two components. Splines can exhibit discontinuities, either in themselves or in their derivatives, at various knot locations, which means that the knots must also be included in the creation of boundary sub-segments.

In summary, the boundary must be divided into segments determined by the nature of the conditions prescribed on it, the points of singularity of the integrands, any possible corners, and any knots occurring in the spline approximation. Once all this information is listed in the `Exclusions` option, the `NIntegrate` function automatically performs the segmentation of the boundary.

4.11 Construction of a B-Spline Basis

Suppose that the boundary ∂S has been partitioned into segments in the manner indicated above. We are using B-splines to represent the two components of the unknown densities φ and ψ that approximate $\mathrm{T}u$ and u, respectively, in the Somigliana formula. If we have mixed boundary conditions, then φ and ψ have support only on the portion of the boundary associated with its specific type of condition. As described earlier, we expect ψ to be continuous on the boundary but to have discontinuous derivatives at the corners. Also, we expect φ to be discontinuous at the corners, where the unit normal vector is undefined.

The splines are constructed in the usual way as piecewise polynomial functions over a domain $0 \leq t \leq \mathrm{tMax}$. A knot t_i is the location between any two adjacent polynomial segments. At each knot, a smoothness condition can be imposed on the adjacent polynomials. It is not necessary to have the same smoothness condition at every knot. In fact, in our applications, we typically have lower smoothness conditions at the knots associated with boundary corners than at the other knots. We refer to the values t_i corresponding to the points that separate the segments $(\partial S)_i$ as primary knots. All the other knots, lying between the primary ones, are called secondary knots and are denoted by $t_{i,j}$. The space between the primary knots is determined by the parameterization of each $(\partial S)_i$. The space between consecutive secondary knots is assumed to be equal on each individual $(\partial S)_i$, but not necessarily the same for different $(\partial S)_i$.

4.28 Example. Consider a piecewise linear spline defined on the interval $0 \leq t \leq 1$ with knots at $0, 1/3, 2/3, 1$. In *Mathematica*®, specification of the knots requires both specification of their location and of the degree of smoothness to be imposed on adjacent polynomials. The function that generates the B-spline basis is

$$\mathrm{BSplineBasis}[\{\mathrm{degree}, \mathrm{knots}\}, i, x], \quad i = 0, \ldots, k.$$

Here, the degree is 1, and the knots are defined with location and smoothness by the set

$\left\{0, 0, \frac{1}{3}, \frac{2}{3}, 1, 1\right\}.$

Figure 4.19 shows the graphs of the four piecewise linear basis functions.

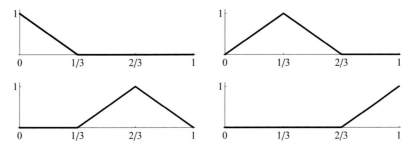

Fig. 4.19 Piecewise linear basis functions.

4.29 Example. Consider piecewise cubic splines that are three times continuously differentiable at the secondary knots, generated with degree 3 and knot locations and smoothness conditions represented by the set

$\left\{0, 0, 0, 0, \frac{1}{3}, \frac{2}{3}, 1, 1, 1, 1\right\}.$

The graphs of the 6 basis functions are shown in Fig. 4.20.

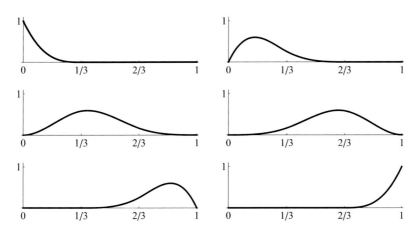

Fig. 4.20 The basis functions for a piecewise smooth cubic B-spline.

4.30 Example. We now consider a piecewise cubic spline that is continuous but with discontinuous first-order derivatives at each knot. Its basis is generated with degree 3 and knot locations and smoothness conditions given by the set

$$\{0, 0, 0, 0, \tfrac{1}{3}, \tfrac{1}{3}, \tfrac{1}{3}, \tfrac{2}{3}, \tfrac{2}{3}, \tfrac{2}{3}, 1, 1, 1, 1\}. \tag{4.22}$$

The graphs of the corresponding 10 basis elements are shown in Fig. 4.21.

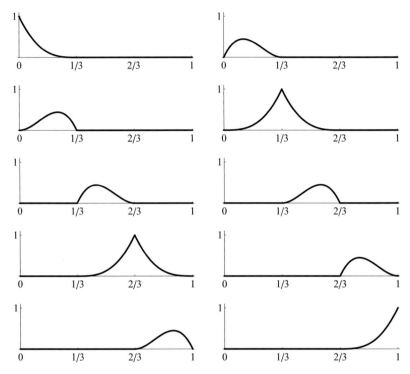

Fig. 4.21 The basis functions of a piecewise continuous (with discontinuous derivatives) cubic B-spline.

4.31 Remark. The bookkeeping for specifying the smoothness condition at a knot may appear to be a little strange, but it does have a logical explanation. A B-spline basis function is defined to be 0 outside the interval determined by its knots. Thus, in Example 4.30, each basis function is 0 for $t < 0$ and $t > 1$. The smoothness condition number m specifies the level of smoothness at the point of contact of adjacent polynomials and does not have to be the same at all the knots. To be specific, $m = 0$ corresponds to discontinuity of the function, $m = 1$ corresponds to continuity of the function and discontinuity of its first-order derivative, $m = 2$ corresponds to continuity of the function and its first-order derivative and discontinuity of its second-order derivative, and so on. The smoothness condition number is computed from the notation for the individual knots by means of the formula

$$m = (\text{degree} + 1) - (n \, \text{knot repeats}),$$

where 'degree' is the degree of the adjacent polynomials and n is the number of times the knot location is repeated in the knot list.

4.32 Example. For the piecewise linear spline considered in Example 4.28, we have degree $= 1$ and $n = 1$; that is, each of the interior secondary knots $1/3$ and $2/3$ is listed only once. Therefore, $m = 1$, which means that the adjacent polynomial basis functions are continuous but have discontinuous first-order derivatives at these knot locations.

4.33 Example. The piecewise cubic spline in Example 4.29 has degree $= 3$ and $n = 1$ at the knots $1/3$ and $2/3$, which yields $m = 3$. This means that the adjacent polynomials at these knots are continuous with continuous first-order and second-order derivatives, but discontinuous third-order derivatives.

4.34 Example. In Example 4.30, we have degree $= 3$ and $n = 3$ at the knots $1/3$ and $2/3$, so $m = 1$, indicating that the adjacent cubic polynomials are continuous but their first-order derivatives are discontinuous at the two interior knots.

4.35 Remark. The starting and stopping knot locations require special treatment. As already mentioned, a spline is defined to be 0 before the first knot and after the last knot. In Example 4.34, each of these knots are repeated four times, which produces the smoothness condition number $m = 0$. Consequently, the first and last basis functions have a jump discontinuity between $0+$ and $1-$. The smoothness condition number for the second and second-last splines is $m = 1$, so each of them starts and stops with the same value (in this case, 0) but with different values for their first-order derivatives. Continuing this pattern, the smoothness condition number m steadily increases at the starting and stopping knot locations until it becomes numerically equal to the polynomial degree. This behavior is illustrated in Figs. 4.19–4.21.

We now turn our attention to the unknown densities φ and ψ on the boundary. They are approximated by

$$\varphi[t] = \varphi[y[t]] \cong b[t], \quad 0 \le t \le \text{tMax},$$
$$\psi[t] = \psi[y[t]] \cong b[t], \quad 0 \le t \le \text{tMax},$$

where b is a B-spline representing an approximation with primary knots t_i determined by all the boundary segments $(\partial S)_i$. Any expected smoothness in φ and ψ should be reflected by the primary knot locations t_i and the smoothness at these locations. As mentioned earlier, it is computationally unacceptable to assume a higher level of smoothness in an approximation than the actual level exhibited by what is being approximated. If the boundary has corners, then there will be discontinuities in the densities themselves or in their derivatives. This, too, should be reflected in the B-spline primary knot locations and the smoothness conditions at these locations.

The equally-spaced secondary knots $t_{i,j}$ and their corresponding piecewise polynomials are usually chosen to achieve the approximation accuracy desired in the problem on each boundary segment $(\partial S)_i$. We recall that the density is assumed to be analytic on each $(\partial S)_i$. Therefore, both the degree of the piecewise polynomials and the smoothness at the secondary knots is dictated strictly by computational and accuracy considerations.

4.36 Example. Let h be the spacing between adjacent secondary knots $t_{i,j}$. Then the expected optimal accuracy for a piecewise linear B-spline approximation is $O(h^2)$, and the expected optimal accuracy for a piecewise cubic B-spline approximation is $O(h^4)$. Whether or not this optimal accuracy can be achieved depends on the specifics of the selected method.

In conclusion, the bookkeeping for constructing a B-spline from a B-spline basis consists of five steps.

Step 1. The boundary ∂S is divided into segments $(\partial S)_i$ with primary knots t_i, $i = 1, \ldots, m$.

Step 2. Each segment $(\partial S)_i$ is subdivided by means of secondary equally-spaced knots $t_{i,j}$.

Step 3. The smoothness at each primary knot t_i is chosen to correspond to the smoothness of φ or ψ.

Step 4. The polynomial degree and smoothness are chosen for the secondary knots, thus defining the piecewise polynomial spline elements on each $(\partial S)_i$.

Step 5. The basis elements $b_{i,j}$ are constructed for each of the two components of φ or ψ.

The procedure for generating the exact basis elements is somewhat complicated to describe and is best illustrated by a few specific cases.

4.37 Example. Consider the half disc domain discussed in Example 4.26, with the boundary, corners, and exclusions shown in Fig. 4.18. The parametrization of the two components $\psi_\alpha[y[t]]$ of the unknown density $\psi[y[t]]$, which represents $u[y[t]]$, has two corners, so it produces three primary knots $t_i = 0, 1, 2$, where $\psi[y[t]]$ is expected to be continuous but with a discontinuous first-order derivative. We introduce two secondary knots in each of the two boundary segments $(\partial S)_1$ and $(\partial S)_2$, and create piecewise cubic splines that are twice continuously differentiable at these knots. Thus, the knot locations and smoothness conditions are given by the set

$$\left\{0, 0, 0, 0, \tfrac{1}{3}, \tfrac{2}{3}, 1, 1, 1, \tfrac{4}{3}, \tfrac{5}{3}, 2, 2, 2, 2\right\}.$$

Figure 4.22 shows the 11 piecewise cubic B-spline basis elements $b_{i,j}$ constructed in this way.

In reality, the density ψ has periodic smoothness between the first and last basis functions rather than the discontinuity indicated above. Consequently, the first basis function needs to be modified to meet this smoothness condition, leaving only 10 of the $b_{i,j}$, as shown in Fig. 4.23.

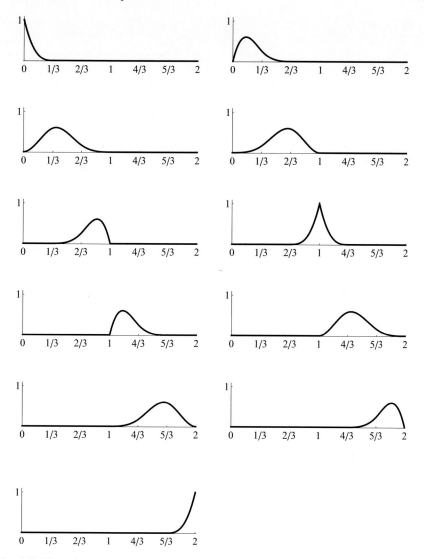

Fig. 4.22 Piecewise continuous, with discontinuous first-order derivatives, cubic (non-periodic) basis functions.

4.38 Example. With the same setup as in Example 4.37, we can also approximate the unknown density φ, which represents Tu. Proceeding as above, we conclude that here the knot locations and smoothness conditions are specified by the set

$$\left\{0, 0, 0, 0, \tfrac{1}{3}, \tfrac{2}{3}, 1, 1, 1, 1, \tfrac{4}{3}, \tfrac{5}{3}, 2, 2, 2, 2\right\}.$$

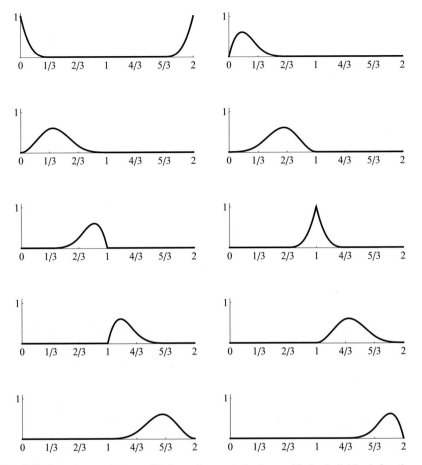

Fig. 4.23 Piecewise continuous, with discontinuous derivatives, cubic (periodic) basis functions.

Figure 4.24 shows the 12 piecewise cubic B-spline basis elements $b_{i,j}$ generated in this case.

The graphs of these functions, restricted to the boundary and written as $b_{i,j}[y[t]]$, are displayed in Fig. 4.25.

After the appropriate B-spline basis functions $b_{i,j}$ have been constructed, we can define the approximate densities $\tilde{\psi}$ and $\tilde{\varphi}$ as

$$\tilde{\psi}[t] = \{\tilde{\psi}_1[y[t]], \tilde{\psi}_2[y[t]]\}$$

$$= \left\{ \sum_i \sum_j c_{1,i,j} b_{i,j}[t], \sum_i \sum_j c_{2,i,j} b_{i,j}[t] \right\}, \quad 0 \le t \le \text{tMax},$$

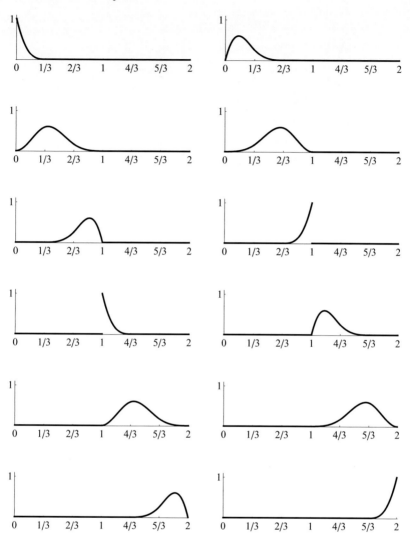

Fig. 4.24 Cubic B-spline basis functions for approximating φ.

and

$$\tilde{\varphi}[t] = \{\tilde{\varphi}_1[y[t]], \tilde{\varphi}_2[y[t]]\}$$

$$= \left\{ \sum_i \sum_j c_{1,i,j} b_{i,j}[t], \sum_i \sum_j c_{2,i,j} b_{i,j}[t] \right\}, \quad 0 \le t \le \text{tMax}.$$

A boundary element method is then applied to solve for the unknown numerical coefficients $c_{\alpha,i,j}$.

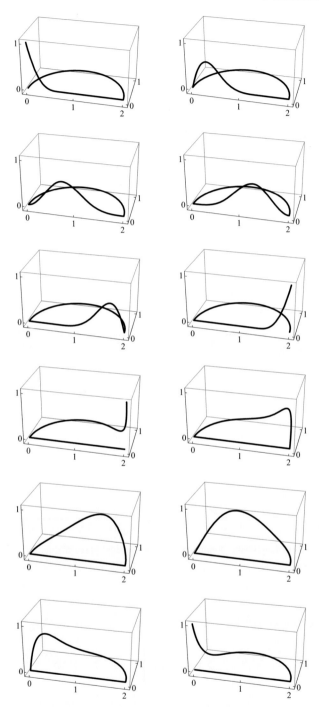

Fig. 4.25 Cubic B-spline basis functions in 3-D.

4.12 Selection of the Boundary Element Method and Collocation Points

The boundary integral method can be combined with the B-spline basis elements $b_{i,j}$ to produce a boundary element method.

4.39 Example. Consider the direct method for the interior Dirichlet problem, which leads to equation $(\mathscr{D}_D^+)$, combined with the B-spline basis elements developed in the previous section; that is,

$$
\left(\oint_{\Gamma_{\mathrm{Weak}[x]}} D[x,y[t]] \circ \begin{pmatrix} \dots & b_{i,j}[t] & \dots & 0 & 0 & 0 \\ 0 & 0 & 0 & \dots & b_{i,j}[t] & \dots \end{pmatrix} \mathrm{dsdt}[t]\, dt \right) \circ \begin{pmatrix} \vdots \\ c_{1,i,j} \\ \vdots \\ c_{2,i,j} \\ \vdots \end{pmatrix}
$$

$$
= \left(\oint_{\Gamma_{\mathrm{CPV}[x]}} P[x,y[t]] \circ \mathscr{P}[y[t]] \mathrm{dsdt}[t]\, dt + \tfrac{1}{2}\,\mathscr{P}[x] \right).
$$

This system generates an approximation $\tilde{\varphi}$ to the actual density φ by means of the formula

$$
\varphi[t] \cong \tilde{\varphi}[t] = \begin{pmatrix} \tilde{\varphi}_1[y[t]] \\ \tilde{\varphi}_2[y[t]] \end{pmatrix} = \begin{pmatrix} \sum_i \sum_j c_{1,i,j} b_{i,j}[t] \\ \sum_i \sum_j c_{2,i,j} b_{i,j}[t] \end{pmatrix}, \quad 0 \le t \le \mathrm{tMax}.
$$

Writing equation $(\mathscr{D}_D^+)$ for φ in the form

$$
V_0(\varphi)(x) - \left[W_0(\mathscr{P})(x) + \tfrac{1}{2}\,\mathscr{P}(x) \right] = 0, \quad x \in \partial S,
$$

and substituting $\tilde{\varphi}$ in it, we define the residual

$$
R(x) = V_0(\tilde{\varphi})(x) - \left[W_0(\mathscr{P})(x) + \tfrac{1}{2}\,\mathscr{P}(x) \right], \quad x \in \partial S,
$$

which is only approximately 0; that is,

$$
R(x) \cong 0, \quad x \in \partial S.
$$

We must now select a strategy for computing the unknown coefficients $c_{\alpha,i,j}$. The usual procedure is to implement a method that makes the residual $R(x)$ small. This can be accomplished in many different ways, each technique having its advantages and disadvantages. Our preference is known as the weighted residual method, which we briefly outline below.

We start by considering a set of linearly independent functions

$$
\{w_1(x), w_2(x), \dots, w_n(x)\}, \quad x \in \partial S, \tag{4.23}
$$

that are sufficiently 'well behaved' to allow the evaluation of the integrals

$$\oint_{\partial S} R(x) w_i(x) \, ds, \quad i = 1, 2, \ldots, n.$$

The method of weighted residuals determines the $c_{\alpha,i,j}$ by requiring $R(x)$ to be orthogonal (with respect to the L^2 inner product) to

$$\text{span} \left\{ w_1(x), w_2(x), \ldots, w_n(x) \right\}.$$

Thus, we want to have

$$\oint_{\partial S} R(x) w_i(x) \, ds \cong 0, \quad i = 1, 2, \ldots, n. \tag{4.24}$$

These conditions generate n equations which, for the method to be successful, should form a consistent algebraic system. If n equals the number of unknown coefficients $c_{\alpha,i,j}$, we may be able to find a unique solution. If the system is overdetermined, then we need to seek a least squares fit to the unknown coefficients.

The next step is to select the functions (4.23), which immediately raises two issues.

Issue 1. In the boundary element method, the residual $R(x)$ is usually very expensive to evaluate numerically. If the w_i have significant nonzero support on ∂S, then the numerical evaluation of the equations generated by (4.24) is also very expensive.

Issue 2. The overall accuracy of the boundary element method depends on a number of factors, including

   (i)   the choice of boundary integral method;
   (ii) the approximating properties of the chosen B-spline basis;
   (iii) the approximating properties of the chosen functions w_i;
   (iv) the choice of weighted residual method;
   (v)   the accuracy of the numerical methods used in the implementation.

Issue 2 indicates that there can be a massive number of possible variations to consider. One popular choice, known as the Galerkin method, selects the weighting functions w_i to be the same as the approximating basis, which in our case is a B-spline basis. This partially addresses issue 1 because the basis has been constructed to have minimal nonzero support, thus simplifying the evaluation of the inner products. The Galerkin method with B-splines also has very good approximating properties, which answers issues 2(ii) and 2(iii).

However, issue 1 is the most important for our applications. Again, the expensive numerical evaluation of $R(x)$ dictates the choice of the w_i. Since the Dirac delta distribution $\delta(x_i, x)$ with $x_i \in \partial S$ minimizes the evaluation of $R(x)$, we make this choice for the w_i, so the weighted residual method reduces to

$$\oint_{\partial S} R(x)\delta(x_i,x)\,ds = 0, \quad i = 1,2,\ldots,n,$$

or

$$R(x_i) = 0, \quad i = 1,2,\ldots,n.$$

This procedure is known as the collocation method and is widely used in practice.

Issue 2(iii), concerning the approximating properties of the $\delta(x_i,x)$, involves both the number and location of the x_i on ∂S. It might appear that the x_i can be chosen without much restriction, but this is not the case. The selected points must, at a minimum, meet the following constraints:

(a) They must ensure that $R(x_i)$ is both uniquely defined and 'well behaved'. Consequently, it is usually unacceptable to have the x_i located at a boundary corner, where the unit normal vector is undefined and the method potentially breaks down. There are procedures to work around this situation, but they require special consideration.

(b) The choice of the x_i must guarantee sufficient independent constraints to determine the $c_{\alpha,i,j}$. For example, the B-spline basis is composed of functions $b_{i,j}$ constructed with minimal support on the boundary. These functions have only a limited amount of smoothness linking adjacent polynomials. As a consequence, it is possible to make a selection of the x_i that does not yield all the coefficients although the total number of these points would seem to be sufficient.

(c) The number and location of the x_i also influence the accuracy of the overall method. For example, their location affects the region of the boundary where $R(x)$ is usually the smallest and, hence, the overall approximation accuracy. One possible solution to consider is selecting the x_i to be the Gaussian quadrature points on each interval between adjacent secondary knots. Another is to select an excess number of collocation points x_i. In practice, this choice often helps increase the accuracy, but has the disadvantage that it generates an overdetermined system, which increases the total computation time.

4.40 Example. We illustrate this in the case considered in the Examples 4.26 and 4.38, where we parameterized the upper half of a circle of radius one. There we stated that we must use the two corners of the boundary to determine the primary knots $t_i = 0, 1, 2$, and that the corresponding smoothness conditions depend on the approximate density $\varphi[y[t]]$. The latter is expected to be discontinuous at the primary knots. We construct a B-spline basis by adjoining secondary knots. In Example 4.38, this was done by placing two secondary knots $t_{i,j}$ between consecutive primary knots, raising the total number of knots to 6. We then constructed a B-spline basis using cubic polynomials with continuous second-order derivatives at the secondary knots and no continuity at the primary knots. This basis, which consists of 12 elements, is shown in Fig. 4.24. The primary and secondary knots are marked on the graph on the left in Fig. 4.26.

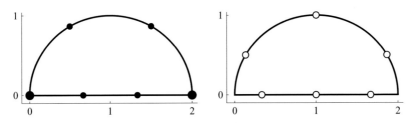

Fig. 4.26 Left: the knot locations on ∂S. Right: the collocation points.

We can select collocation points $x_i \in \partial S$ to determine the 12 unknown coefficients $c_{\alpha,i,j}$. Because each equation of the boundary element method has two components, we need at least six collocation points. We choose these points midway between adjacent knots. Of course, the choice is not unique, but this particular one provides a good position for the collocation points since it makes the integrand $D[x_i, y[t]] \circ \tilde{\varphi}[y[t]]$ 'well behaved' in the vicinity of x_i. Our 6 collocation points are marked on the graph on the right in Fig. 4.26.

4.13 Code Validation

The four primary steps of this essential process in the construction of the code for our methods are

(i)   theoretical code validation;
(ii) visualization of the code;
(iii) validation of both small and large code parts;
(iv) validation with known test examples.

Each step of the code development implements a specific portion of the theory for solving our boundary integral equations, which means that we can often make a theoretical check when the step has been completed. For example, as seen above, computing D and P generates a large amount of symbolic code. It is known that these matrices must satisfy (4.6) and (4.10). Frequently, this can be verified by direct substitution and use of the symbolic simplification capabilities of *Mathematica*®. It should be understood that the functions `Simplify` and `FullSimplify` return the primary generic result of a simplification, excluding such issues as singular points. These functions do not locate the singularities and do not generate $\delta[x,y]$. Both D and P are extremely large, which makes $Z_x \diamond D[x,y]$ and $Z_x \diamond P[x,y]$ even larger. Consequently, `Simplify` and `FullSimplify` can take a substantial amount of time to reach the desired result. An explanation of the scope of these two functions was given in Remark 4.15.

As a rule, it is desirable to have numerical validation in addition to symbolic theoretical validation. The best way to validate a substantial amount of numerical calculations is through visualization. Sometimes, however, this is not practical, but in most circumstances it can be readily enacted.

4.41 Example. Visualizing the requirement that D and P must satisfy (4.6) and (4.10)—that is,

$$Z_x \diamond D[x,y] = \begin{pmatrix} 0 & 0 \\ 0 & 0 \end{pmatrix},$$

$$Z_x \diamond P[x,y] = \begin{pmatrix} 0 & 0 \\ 0 & 0 \end{pmatrix},$$

is straightforward. A downside of using numerical visualization is that we need to assign specific values to all the parameters in D and P in order to have the symbolic expression reduce to a numerical one, and it is always possible that such specific test values may be either unrepresentative or inappropriate. In our case, Fig. 4.27 illustrates visually the condition for $Z_x \diamond D[x,y]$ with $x = \{0.5, 0.5\}$. These graphs show that each of the four components is 0 except possibly near the singularity. Because a large portion of the boundary integral equation code is numerical in nature, visualization becomes one of the most important aspects of the validation process.

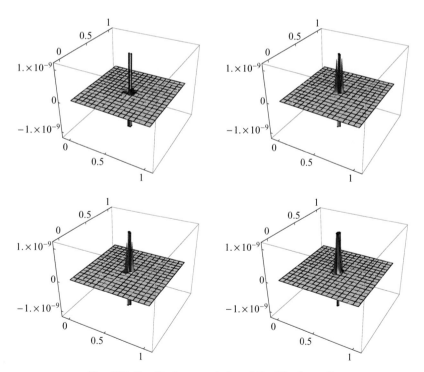

Fig. 4.27 Graphical representation of $Z_x \diamond D[x,y]$, $x \in S$.

It is also important to validate both small and large portions of the code. Validation of small code segments helps ensure that the code has been created properly; validation of larger code segments guarantees that the individual smaller components are working together as required. Small code validation is often easy and straightforward, and can usually be accomplished by both symbolic and numerical procedures. Every effort should be made to implement as many such validation tests as possible since a small segment of code, owing to its brevity, is normally well understood and easy to debug. Delaying debugging to larger code segments is undesirable because it is often unclear what portion of the code is causing the problem. However, validation of large code segments remains important, as it enables us to verify the overall functioning of the program.

The validation process must finish with a number of test cases that implement the full and complete code. For our model, a test case should be a nontrivial solution of the homogeneous problem

$$Z_x \diamond u[x] = \begin{pmatrix} 0 \\ 0 \end{pmatrix}. \tag{4.25}$$

Following a procedure similar to that used to derive the fundamental matrix of solutions, we assume that the solution we are seeking is of the form

$$u[x] = \text{adjoint}[Z_x] \diamond \begin{pmatrix} \text{test}_1[x] \\ \text{test}_2[x] \end{pmatrix}. \tag{4.26}$$

Inserting this expression in the above equation yields

$$Z_x \diamond \text{adjoint}[Z_x] \diamond \begin{pmatrix} \text{test}_1[x] \\ \text{test}_2[x] \end{pmatrix} = \text{Det}[Z_x] \diamond \begin{pmatrix} \text{test}_1[x] \\ \text{test}_2[x] \end{pmatrix} = \begin{pmatrix} 0 \\ 0 \end{pmatrix},$$

so the two components of the solution will be constructed from the solution of the equation

$$\text{Det}[Z_x] \diamond \text{test}_\alpha[x] = 0,$$

which, as shown in Sect. 4.4, is the same as

$$\mu(\lambda + 2\mu) \left(\Delta - \frac{k}{\mu} \mathfrak{I} \right) \left(\Delta - \frac{k}{\lambda + 2\mu} \mathfrak{I} \right) \diamond \text{test}_\alpha[x] = 0.$$

Changing to polar form and continuing the reasoning in that section, we arrive at the linear combinations

$$\text{test}_\alpha[a_\alpha, b_\alpha, r] = a_\alpha \text{BesselI} \left[0, r \sqrt{\frac{k}{\mu}} \right] + b_\alpha \text{BesselI} \left[0, r \sqrt{\frac{k}{\lambda + 2\mu}} \right], \tag{4.27}$$

where, in view of the homogeneous nature of (4.25), we have chosen the nonsingular function BesselI instead of BesselK. Once again, the code can be used to check the accuracy of the above expression by verifying symbolically that

$$\text{Det}\,[Z_x] \diamond \text{test}_\alpha\,[a_\alpha, b_\alpha, x] = 0.$$

All the ingredients are now available to compute the desired test solution (4.26). Using `FullSimplify` and, say, the specific values

$$\{a_1, a_2\} = \{1, -3\}, \quad \{b_1, b_2\} = \{2, 4\},$$

we reduce the two components $u_\alpha[1, 2, -3, 4, \text{x1}, \text{x2}]$ to

$$
\Big\{ \big(k(\lambda + \mu)(2\text{x}2(\lambda + 2\mu)
$$

$$
\text{HypergeometricOF1Regularized}\left[1, \frac{k((1+\text{x}1)^2 + \text{x}2^2)}{4\mu}\right]
$$

$$
(\text{x}2a_1 - (1+\text{x}1)a_2) + (\lambda + 2\mu)
$$

$$
\text{HypergeometricOF1Regularized}\left[2, \frac{k((1+\text{x}1)^2 + \text{x}2^2)}{4\mu}\right]
$$

$$
((1+\text{x}1 - \text{x}2)(1+\text{x}1+\text{x}2)a_1 + 2(1+\text{x}1)\text{x}2a_2) - 2(1+\text{x}1)\mu
$$

$$
\text{HypergeometricOF1Regularized}\left[1, \frac{k((1+\text{x}1)^2 + \text{x}2^2)}{4(\lambda + 2\mu)}\right]
$$

$$
((1+\text{x}1)b_1 + \text{x}2b_2) + \mu
$$

$$
\text{HypergeometricOF1Regularized}\left[2, \frac{k((1+\text{x}1)^2 + \text{x}2^2)}{4(\lambda + 2\mu)}\right]
$$

$$
((1+\text{x}1 - \text{x}2)(1+\text{x}1+\text{x}2)b_1 + 2(1+\text{x}1)\text{x}2b_2))) /
$$

$$
(2((1+\text{x}1)^2 + \text{x}2^2)\mu(\lambda + 2\mu)),
$$

$$
(k(\lambda + \mu)(2(1+\text{x}1)(\lambda + 2\mu)
$$

$$
\text{HypergeometricOF1Regularized}\left[1, \frac{k((1+\text{x}1)^2 + \text{x}2^2)}{4\mu}\right]
$$

$$
(-\text{x}2a_1 + (1+\text{x}1)a_2) + (\lambda + 2\mu)
$$

$$
\text{HypergeometricOF1Regularized}\left[2, \frac{k((1+\text{x}1)^2 + \text{x}2^2)}{4\mu}\right]
$$

$$
(2(1+\text{x}1)\text{x}2a_1 + (-(1+\text{x}1)^2 + \text{x}2^2)a_2) - 2\text{x}2\mu
$$

$$
\text{HypergeometricOF1Regularized}\left[1, \frac{k((1+\text{x}1)^2 + \text{x}2^2)}{4(\lambda + 2\mu)}\right]
$$

$$
((1+\text{x}1)b_1 + \text{x}2b_2) + \mu
$$

$$
\text{HypergeometricOF1Regularized}\left[2, \frac{k((1+\text{x}1)^2 + \text{x}2^2)}{4(\lambda + 2\mu)}\right]
$$

$$
(2(1+\text{x}1)\text{x}2b_1 + (-(1+\text{x}1)^2 + \text{x}2^2)b_2))) /
$$

$$
(2((1+\text{x}1)^2 + \text{x}2^2)\mu(\lambda + 2\mu)) \Big\}.
$$

This test solution can be verified symbolically by applying `FullSimplify` to check that it satisfies (4.25). The verification can also be done graphically, as illustrated in Fig. 4.28.

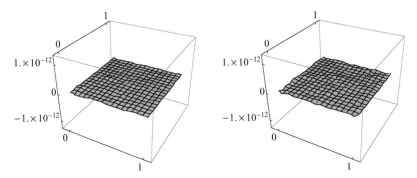

Fig. 4.28 Graphs of the evaluation of the components of $Z_x \diamond u[a_1, \beta_1, \alpha_2, \beta_2, x]$.

Full verification of the code can be achieved by considering a wide range of particular test cases.

4.42 Example. Consider the homogeneous equation in S with a Dirichlet condition on ∂S; that is,

$$Z_x \diamond u[x] = Z_x \diamond \begin{pmatrix} u_1[x] \\ u_2[x] \end{pmatrix} = \begin{pmatrix} 0 \\ 0 \end{pmatrix}, \quad x \in S,$$

$$u[x] = \begin{pmatrix} u_1[x] \\ u_2[x] \end{pmatrix} = \begin{pmatrix} \mathscr{P}_1[x] \\ \mathscr{P}_2[x] \end{pmatrix}, \quad x \in \partial S.$$

The data function $\mathscr{P}$ is constructed from a known test solution $u[a_1, b_1, a_2, b_2, x]$ restricted to the boundary. We use the test solution generated with the parameter choice

$$\lambda \to 1, \quad \mu \to 2, \quad k \to 3,$$
$$\alpha_1 \to 1, \quad \beta_1 \to 2, \quad \alpha_2 \to -3, \quad \beta_2 \to 4.$$

Our test domain S is the half disc featured in Example 4.26 (see Figure 4.18), with the same parameterization (4.21) of the boundary ∂S. The graphs of the two components of the test solution u in S are shown in Fig. 4.29.

The two components of $\mathscr{P}$ on ∂S can be constructed directly from $u[x]$ by restricting x to the boundary:

$$\mathscr{P}[x] = u[x], \quad x \in \partial S.$$

The graphs of the components of this function are displayed in Fig. 4.30.

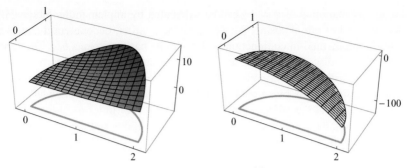

Fig. 4.29 The components of $u[x]$, $x \in S$.

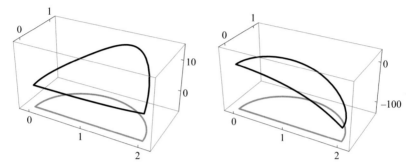

Fig. 4.30 The components of $\mathscr{P}$.

We solve our Dirichlet problem by means of the direct method described in Sect. 3.2, which yields the boundary equation $(\mathscr{D}_D^+)$ (see Example 4.24). This equation is coded in the form (4.19), where the unknown density φ represents Tu. Once φ is determined, we find the solution in S from the representation formula (4.20).

We can use our test solution u to construct the exact function Tu by evaluating

$$(\text{Tu})[x] = T_x \diamond u[x] = T_x \diamond \begin{pmatrix} u_1[x] \\ u_2[x] \end{pmatrix} = \begin{pmatrix} \text{Tu}_1[x] \\ \text{Tu}_2[x] \end{pmatrix}, \quad x \in \partial S.$$

The graphs of both components of Tu are shown in Fig. 4.31.

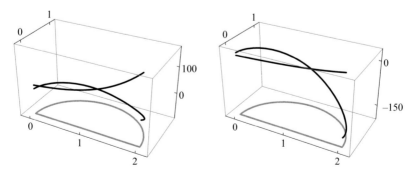

Fig. 4.31 The components of Tu.

An approximation $\tilde{\varphi}$ of $\varphi = \mathrm{Tu}$ can be calculated by implementing a boundary element method of our choosing. As we mentioned earlier, our preferred choice is the collocation method in conjunction with B-spline basis elements $b_{i,j}$. The function φ will have discontinuities at the two domain corners, and the approximating B-spline basis will of necessity reflect this feature. For our problem, we use a piecewise cubic spline that is twice continuously differentiable at the secondary knots and discontinuous at the corner knot locations $t = 0, 1, 2$. In terms of the selected parameterization, the primary and secondary knot locations and the corresponding smoothness conditions are specified by the set

$$\left\{ 0, 0, 0, 0, \tfrac{1}{4}, \tfrac{1}{2}, \tfrac{3}{4}, 1, 1, 1, 1, \tfrac{5}{4}, \tfrac{3}{2}, \tfrac{7}{4}, 2, 2, 2, 2 \right\}.$$

These locations, marked on the graph on the left in Fig. 4.32, determine 14 functions $b_{i,j}$ for each component of $\tilde{\varphi}$. The graphs of these functions are shown in parametric form in Fig. 4.33.

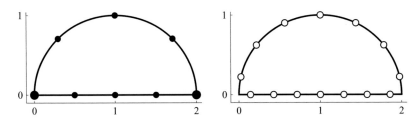

Fig. 4.32 Left: the knot locations on ∂S. Right: the collocation points.

We need to choose a suitable number of collocation points to be able to compute a solution $\tilde{\varphi}$. This can be done in several different ways. But we must remember that a collocation point cannot be placed at any of the corners $t = 0, 1, 2$, where φ is expected to be discontinuous. The 14 parametric values for t used as collocation points and displayed on the graph on the right in Fig. 4.32 are

$$\left\{ \tfrac{1}{14}, \tfrac{3}{14}, \tfrac{5}{14}, \tfrac{1}{2}, \tfrac{9}{14}, \tfrac{11}{14}, \tfrac{13}{14}, \tfrac{15}{14}, \tfrac{17}{14}, \tfrac{19}{14}, \tfrac{3}{2}, \tfrac{23}{14}, \tfrac{25}{14}, \tfrac{27}{14} \right\}.$$

The same 14 collocation points are used for both components of $\tilde{\varphi}$ to produce 28 equations, from which we compute the coefficients $c_{\alpha,i,j}$ in the representation

$$\tilde{\varphi}[t] = \{ \tilde{\varphi}_1[x[t]], \tilde{\varphi}_2[x[t]] \}$$

$$= \left\{ \sum_i \sum_j c_{1,i,j} b_{i,j}[t], \sum_i \sum_j c_{2,i,j} b_{i,j}[t] \right\}, \quad 0 \le t \le \mathrm{tMax}.$$

The graphs of the two components of $\tilde{\varphi}$ are shown in Fig. 4.34. This approximate solution agrees very well with the exact solution $\varphi = \mathrm{Tu}$ (see Fig. 4.31).

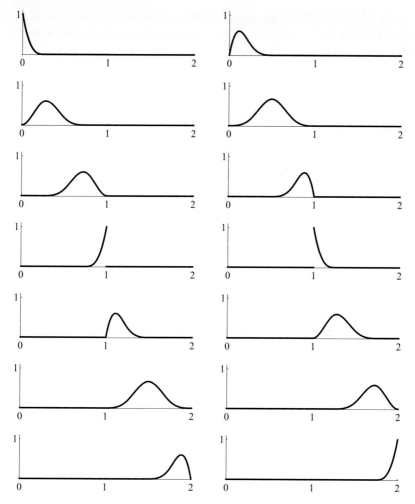

Fig. 4.33 The 14 B-spline basis functions $b_{i,j}[t]$, $0 \le t \le 2$.

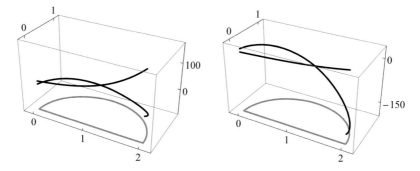

Fig. 4.34 The components of $\tilde{\varphi}$.

To get a better understanding of the discrepancy between $\tilde{\varphi}$ and φ, in Fig.4.35 we have graphed the difference $\tilde{\varphi} - \varphi$ relative to the maximum absolute value of φ on ∂S.

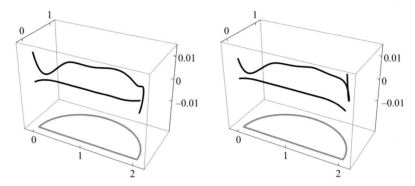

Fig. 4.35 The components of the relative error in $\tilde{\varphi}$ (Cartesian coordinates).

Since Fig. 4.35 is somewhat difficult to interpret, in Fig. 4.36 we have graphed the same function in parametric form for $0 \leq t \leq 2$. These new graphs indicate that the relative error in the two components of $\tilde{\varphi}$ is only about 0.2%, except possibly near the corners of ∂S.

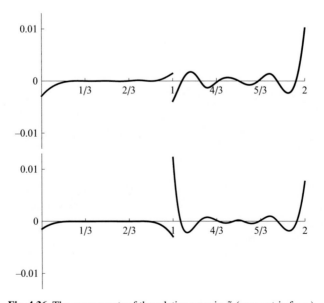

Fig. 4.36 The components of the relative error in $\tilde{\varphi}$ (parametric form).

We can now use $\tilde{\varphi}$ and $\mathscr{P}$ together with the representation formula

$$\tilde{u}[x] = \oint_\Gamma D[x,y] \circ \tilde{\varphi}[y] \, d\Gamma_y - \oint_\Gamma P[x,y] \circ u[y] \, d\Gamma_y, \quad x \in S,$$

to compute an approximation $\tilde{u}$ of the exact solution u in S. The graphs of the two components of $\tilde{u}$ are displayed in Fig. 4.37. Comparing these graphs with those of the components of u in Fig. 4.29, we conclude that the difference between $\tilde{u}$ and u cannot be discerned visually.

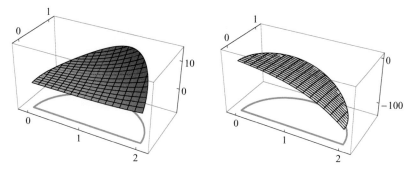

Fig. 4.37 The components of $\tilde{u}$ in S.

4.43 Remark. Example 4.42 and many others presented in the next chapter for all types of boundary conditions and boundary integral methods validate the computer code of our *Mathematica*® program.

Chapter 5
Computational Examples

5.1 Preliminaries

5.1.1 Test Solution

As established in the earlier part of the book, the equilibrium system governing our mathematical model is

$$Z_x \diamond u[x] = 0, \quad x \in S, \tag{5.1}$$

where u is the displacement vector function and Z_x is the operator defined by (4.1). The examples discussed in this chapter illustrate the numerical implementation of the direct and classical indirect methods for various boundary value problems. Since it is important to know how accurate our results are, in a majority of cases we make use, for comparison purposes, of a test solution of system (5.1). Specifically, using (4.26) and (4.27) with parameters

$$\lambda \to 1, \quad \mu \to 2, \quad k \to 3, \quad a_1 = 1, \quad b_1 = 2, \quad a_2 = -3, \quad b_2 = 4 \tag{5.2}$$

in conjunction with the *Mathematica*® function FullSimplify, we construct the particular solution

$$u[x1, x2]$$

$$= \left\{ \frac{9}{4} \left(\left(2\,x2^2 \text{Hypergeometric0F1}\left[1, \tfrac{3}{8}\left((1+x1)^2 + x2^2\right)\right] \right. \right.$$

$$\left. + (1+x1-x2)(1+x1+x2)\text{Hypergeometric0F1}\left[2, \tfrac{3}{8}\left((1+x1)^2 + x2^2\right)\right] \right) \Big/$$

$$\left((1+x1)^2 + x2^2\right)$$

$$+ \frac{4}{5}\left(-\left(2(1+x1)^2 \text{BesselI}\left[2, \sqrt{\tfrac{3}{5}}\sqrt{(1+x1)^2 + x2^2}\right] \right) \Big/ \left((1+x1)^2 + x2^2\right) \right.$$

$$\left. - \text{Hypergeometric0F1Regularized}\left[2, \tfrac{3}{20}\left((1+x1)^2 + x2^2\right)\right] \right)$$

© Springer International Publishing Switzerland 2016
C. Constanda et al., *Boundary Integral Equation Methods and Numerical Solutions*,
Developments in Mathematics 35, DOI 10.1007/978-3-319-26309-0_5

$$-\tfrac{12}{25}(1+x1)x2\,\mathrm{HypergeometricOF1Regularized}\big[3,\tfrac{3}{20}\big((1+x1)^2+x2^2\big)\big]$$

$$+\tfrac{9}{4}(1+x1)x2\,\mathrm{HypergeometricOF1Regularized}\big[3,\tfrac{3}{8}\big((1+x1)^2+x2^2\big)\big]\Big),$$

$$\tfrac{9}{4}\Big(8\Big(-2\,x2^2\mathrm{HypergeometricOF1Regularized}\big[1,\tfrac{3}{20}\big((1+x1)^2+x2^2\big)\big]$$

$$+\big(-(1+x1)^2+x2^2\big)$$

$$\mathrm{HypergeometricOF1Regularized}\big[2,\tfrac{3}{20}\big((1+x1)^2+x2^2\big)\big]\Big)\Big/$$

$$\Big(5\big((1+x1)^2+x2^2\big)\Big)$$

$$-\tfrac{6}{25}(1+x1)x2\,\mathrm{HypergeometricOF1Regularized}\big[3,\tfrac{3}{20}\big((1+x1)^2+x2^2\big)\big]$$

$$-\tfrac{3}{4}(1+x1)x2\,\mathrm{HypergeometricOF1Regularized}\big[3,\tfrac{3}{8}\big((1+x1)^2+x2^2\big)\big]$$

$$-\tfrac{3}{4}\Big(4\,\mathrm{HypergeometricOF1Regularized}\big[2,\tfrac{3}{8}\big((1+x1)^2+x2^2\big)\big]$$

$$+3(1+x1)^2\mathrm{HypergeometricOF1Regularized}\big[3,\tfrac{3}{8}\big((1+x1)^2+x2^2\big)\big]\Big)\Big)\Big\}.$$

$$(5.3)$$

After simplification, the corresponding expression for $(Tu)[x]$, constructed with (4.2) and the same parameter values (5.2), is

$$(Tu)[x1,x2]$$

$$=\Big\{9\Big(80\big((1+x1)^3+6(1+x1)^2x2-3(1+x1)x2^2-2\,x2^3\big)$$

$$\mathrm{HypergeometricOF1}\big[1,\tfrac{3}{20}\big((1+x1)^2+x2^2\big)\big]$$

$$+100\big((1+x1)^3-9(1+x1)^2x2-3(1+x1)x2^2+3\,x2^3\big)$$

$$\mathrm{HypergeometricOF1}\big[1,\tfrac{3}{8}\big((1+x1)^2+x2^2\big)\big]$$

$$-2\Big(5(1+x1)^3\big(11+3\,x1(2+x1)\big)+30(1+x1)^2\big(9+x1(2+x1)\big)x2$$

$$+6(1+x1)\big(-17+3\,x1(2+x1)\big)x2^2+4\big(-11+9\,x1(2+x1)\big)x2^3$$

$$+3(1+x1)x2^4+6\,x2^5\Big)\mathrm{HypergeometricOF1}\big[2,\tfrac{3}{20}\big((1+x1)^2+x2^2\big)\big]$$

$$+25\Big(-4(1+x1)^3+9(1+x1)^2\big(5+x1(2+x1)\big)x2+3(1+x1)$$

$$\big(5+x1(2+x1)\big)x2^2+3\big(-1+3\,x1(2+x1)\big)x2^3+3(1+x1)x2^4\Big)$$

$$\mathrm{HypergeometricOF1}\big[2,\tfrac{3}{8}\big((1+x1)^2+x2^2\big)\big]\Big)v1[x1,x2]\Big/$$

$$\Big(50\big((1+x1)^2+x2^2\big)^2\Big)$$

$$+9\Big(-160\big(2(1+x1)^3-3(1+x1)^2x2-6(1+x1)x2^2+x2^3\big)$$

$$\mathrm{HypergeometricOF1Regularized}\big[1,\tfrac{3}{20}\big((1+x1)^2+x2^2\big)\big]$$

$$+200\big(3(1+x1)^3+3(1+x1)^2x2-9(1+x1)x2^2-x2^3\big)$$

$$\mathrm{HypergeometricOF1Regularized}\big[1,\tfrac{3}{8}\big((1+x1)^2+x2^2\big)\big]$$

$$-16\Big(-20(1+x1)^3+3(1+x1)^2\big(11+x1(2+x1)\big)x2+6(1+x1)$$

$$(11+x1(2+x1))x2^2 + (-7+3\,x1(2+x1))x2^3 + 6(1+x1)x2^4\Big)$$

$$\text{HypergeometricOF1Regularized}\Big[2, \tfrac{3}{20}\big((1+x1)^2+x2^2\big)\Big]$$

$$-25\Big(3(1+x1)^3\big(11+3\,x1(2+x1)\big)+3(1+x1)^2\big(9+x1(2+x1)\big)x2$$

$$-72(1+x1)x2^2 - 8\,x2^3 - 9(1+x1)x2^4 - 3\,x2^5\Big)$$

$$\text{HypergeometricOF1Regularized}\Big[2, \tfrac{3}{8}\big((1+x1)^2+x2^2\big)\Big]\Big)v2[x1,x2]\Big/$$

$$\Big(100((1+x1)^2+x2^2)^2\Big),$$

$$9\Big(-160(2(1+x1)^3 - 3(1+x1)^2 x2 - 6(1+x1)x2^2 + x2^3)$$

$$\text{HypergeometricOF1Regularized}\Big[1, \tfrac{3}{20}\big((1+x1)^2+x2^2\big)\Big]$$

$$+200\big(3(1+x1)^3+3(1+x1)^2 x2 - 9(1+x1)x2^2 - x2^3\big)$$

$$\text{HypergeometricOF1Regularized}\Big[1, \tfrac{3}{8}\big((1+x1)^2+x2^2\big)\Big]$$

$$-16\Big(-20(1+x1)^3 + 3(1+x1)^2\big(11+x1(2+x1)\big)x2 + 6(1+x1)$$

$$\big(11+x1(2+x1)\big)x2^2 + \big(-7+3\,x1(2+x1)\big)x2^3 + 6(1+x1)x2^4\Big)$$

$$\text{HypergeometricOF1Regularized}\Big[2, \tfrac{3}{20}\big((1+x1)^2+x2^2\big)\Big]$$

$$-25\Big(3(1+x1)^3\big(11+3\,x1(2+x1)\big)+3(1+x1)^2\big(9+x1(2+x1)\big)x2$$

$$-72(1+x1)x2^2 - 8\,x2^3 - 9(1+x1)x2^4 - 3\,x2^5\Big)$$

$$\text{HypergeometricOF1Regularized}\Big[2, \tfrac{3}{8}\big((1+x1)^2+x2^2\big)\Big]\Big)v1[x1,x2]\Big/$$

$$\Big(100((1+x1)^2+x2^2)^2\Big)$$

$$-9\Big(80((1+x1)^3 + 6(1+x1)^2 x2 - 3(1+x1)x2^2 - 2\,x2^3)$$

$$\text{HypergeometricOF1}\Big[1, \tfrac{3}{20}\big((1+x1)^2+x2^2\big)\Big]$$

$$+100\big((1+x1)^3 - 9(1+x1)^2 x2 - 3(1+x1)x2^2 + 3\,x2^3\big)$$

$$\text{HypergeometricOF1}\Big[1, \tfrac{3}{8}\big((1+x1)^2+x2^2\big)\Big]$$

$$+2\Big((1+x1)^3\big(-37+3\,x1(2+x1)\big)+6(1+x1)^2\big(-39+x1(2+x1)\big)x2$$

$$+6(1+x1)\big(23+3\,x1(2+x1)\big)x2^2$$

$$+4\big(29+9\,x1(2+x1)\big)x2^3 + 15(1+x1)x2^4 + 30\,x2^5\Big)$$

$$\text{HypergeometricOF1}\Big[2, \tfrac{3}{20}\big((1+x1)^2+x2^2\big)\Big]$$

$$+25\Big(-4(1+x1)^3 + 9(1+x1)^2\big(5+x1(2+x1)\big)x2 + 3(1+x1)$$

$$\big(5+x1(2+x1)\big)x2^2 + 3\big(-1+3\,x1(2+x1)\big)x2^3 + 3(1+x1)x2^4\Big)$$

$$\text{HypergeometricOF1}\Big[2, \tfrac{3}{8}\big((1+x1)^2+x2^2\big)\Big]\Big)v2[x1,x2]\Big/$$

$$\Big(50((1+x1)^2+x2^2)^2\Big)\Big\}, \qquad (5.4)$$

where $v1[x1,x2]$ and $v2[x1,x2]$ are the components of the unit normal to ∂S.

5.1 Remark. The expressions of both $u[x]$ and $(\mathrm{Tu})[x]$ are constructed with the functions `Hypergeometric0F1` and `Hypergeometric0F1Regularized`. The former is the confluent hypergeometric function $_0F_1(a;z)$ and the latter is the regularized confluent hypergeometric function $_0F_1(a;z)/\Gamma(a)$ (see the *Mathematica*® documentation).

5.1.2 Computational Accuracy

In Chap. 4, we explained in great detail that the evaluation of singular integrals requires a significant amount of computational attention. As a consequence, the entire code for the examples presented here is written using exact arithmetic. Floating point machine accuracy is usually limited to about 16 digits. However, when handling the singularities occurring in our problems, we often need to compute with 25 or more significant digits. This places a severe restriction on how the code is written. In *Mathematica*®, the accuracy of an expression is always dictated by the accuracy of its least accurate term. Consequently, it is impossible for the entire code to contain floating point numbers with a predefined fixed accuracy.

There is a difference between representational accuracy and computational accuracy. Suppose that the number π is to be entered with a representational accuracy of five digits. The floating point number 3.1415 can be entered with full representational accuracy as 31415/10000. On the other hand, suppose that our computation of a difficult expression would result in a loss of 10 digits. Then, in order to preserve the five digits of representational accuracy for our number, the computational accuracy must be increased to 15 digits.

The examples in this chapter have been computed with a representational accuracy of 10 digits. For the 'better behaved' singularities encountered in them, there is normally a loss of less than five computational digits; hence, for these examples, 16-digit machine accuracy is sufficient. However, in some 'ill behaved' cases, a loss of 10 digits or more is possible, so, since here machine accuracy is not adequate, the computational accuracy must be increased to at least 20 digits.

In *Mathematica*®, the representational accuracy of a computation for a quantity x is controlled by `PrecisionGoal` and `AccuracyGoal`. The distinction between the two is determined by how the error is bounded; that is,

$$|\text{error}| \leq 10^{-\text{AccuracyGoal}} + |x|10^{-\text{PrecisionGoal}}.$$

Computational accuracy is specified by the setting for `WorkingPrecision`, which is defined as the number of digits used in the computation. Normally, the `WorkingPrecision` must exceed the values of both the `PrecisionGoal` and `AccuracyGoal` by 5 to 10 digits; in certain circumstances, this number may be even larger.

The above discussion refers to controlling the loss of accuracy for a single computational step. But the final answer is usually the result of a large number of such steps, each involving a loss of accuracy. This loss will accumulate to produce an answer with less than the desired representational accuracy. So, for example, if five digits of accuracy are desired in the final answer of a procedure where six digits are lost due to the cumulative computation and seven digits may be lost for a single difficult computation, then the WorkingPrecision should be set to at least 18 digits when performing that difficult computation. We emphasize that the largest part of the program produces only a small loss of accuracy, which means that only a limited portion of the code requires a high value of WorkingPrecision. The WorkingPrecision, PrecisionGoal, and AccuracyGoal have limitations, and their proper use requires careful consideration, numerical validation, and continual vigilance.

5.2 Remark. The examples discussed in the rest of this chapter share a common, streamlined template. To avoid forcing the reader to flip back through the pages in search of various quoted formulas, we have made each example self-contained by concentrating the necessary (analytic, numerical, and graphical) information in each individual case and adopting a generic economical style of presentation. After considering various other options, we felt that a certain amount of repetitiveness was a small price to pay to have the examples easily and independently accessible.

5.2 Dirichlet Problem in an Ellipse: Piecewise Linear Spline

5.2.1 Summary

This example illustrates the application of the direct method to the Dirichlet problem in a convex symmetric domain with a smooth boundary curve. A known test solution is used to calculate the corresponding boundary data function. The collocation method is then implemented, with piecewise linear splines on equally spaced collocation points, to construct an approximation of the original solution from its boundary values. The results are validated through a comparison between the exact and approximate solutions.

5.2.2 Problem Statement

Domain boundary. ∂S is the ellipse parameterized by

$$x1[t] = 1 + \mathrm{Cos}[\pi t], \quad x2[t] = \tfrac{1}{2}\mathrm{Sin}[\pi t] + \tfrac{1}{2}, \quad 0 \le t \le 2. \tag{5.5}$$

Its graph can be seen in Fig. 5.6.

Governing equations. The displacement u is the solution of the boundary value problem

$$Z_x \diamond u[x] = 0, \quad x \in S,$$
$$u[x] = \mathscr{P}[x], \quad x \in \partial S.$$

Test solution. The two components of the test solution u given by (5.3) are graphed in Fig. 5.1.

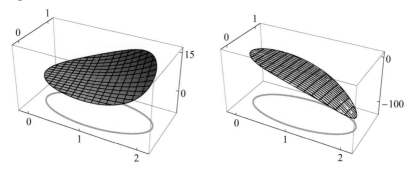

Fig. 5.1 The components of $u[x]$, $x \in S$.

Boundary data function. The function $\mathscr{P}$ is computed from the test solution u as

$$\mathscr{P}[x] = u[x], \quad x \in \partial S.$$

The graphs of its two components are displayed in Figs. 5.2 and 5.3, in Cartesian coordinates and parametric form, respectively.

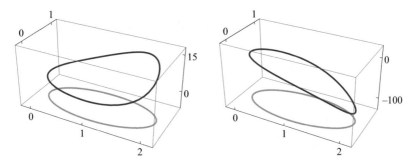

Fig. 5.2 The components of $\mathscr{P}[x]$ (Cartesian coordinates).

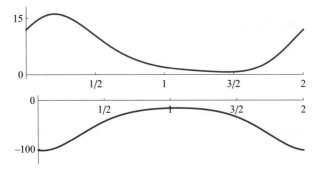

Fig. 5.3 The components of $\mathscr{P}[x[t]]$ (parametric form).

5.2.3 Solution Procedure

Method. We apply the direct method, which reduces the problem to the boundary integral equation

$$V_0(\varphi) = \left(W_0 + \tfrac{1}{2}I\right)\mathscr{P} \quad \text{on } \partial S,$$

coded as

$$\oint_{\Gamma_{\text{Weak}[x]}} D[x,y] \circ \varphi[y]\, d\Gamma_y = \oint_{\Gamma_{\text{CPV}[x]}} P[x,y] \circ \mathscr{P}[y]\, d\Gamma_y + \tfrac{1}{2}\,\mathscr{P}[x]. \tag{5.6}$$

This equation is solved numerically to obtain an approximate density $\tilde{\varphi}$. In turn, $\tilde{\varphi}$ is used in the representation

$$u[x] = \oint_{\Gamma} D[x,y] \circ \varphi[y]\, d\Gamma_y - \oint_{\Gamma} P[x,y] \circ \mathscr{P}[y]\, d\Gamma_y, \quad x \in S \tag{5.7}$$

to generate an approximate solution $\tilde{u}$ in S.

Density function. The exact density φ is the boundary stress vector Tu computed from the test solution u and given by (5.4); its components are graphed in Figs. 5.4 and 5.5.

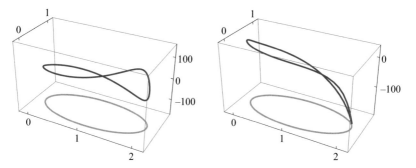

Fig. 5.4 The components of $\varphi[x]$ (Cartesian coordinates).

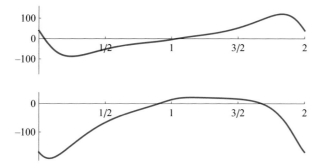

Fig. 5.5 The components of $\varphi[x[t]]$ (parametric form).

Numerical approximation. We compute $\tilde{\varphi}$ by the collocation method with a B-spline basis consisting of elements $b_{i,j}$. Then the approximate density is sought in the form

$$\tilde{\varphi}[x[t]] = \begin{pmatrix} \sum_i \sum_j c_{1,i,j} b_{i,j}[t] \\ \sum_i \sum_j c_{2,i,j} b_{i,j}[t] \end{pmatrix}, \tag{5.8}$$

where the numerical coefficients $c_{\alpha,i,j}$ are determined by substituting (5.8) in equality (5.6).

Since φ is continuous on the boundary, we use a piecewise linear spline with continuity at all knot locations. In terms of the parameterization (5.5), the primary and secondary knots and the smoothness at their locations are specified as the set

$$\{0, \tfrac{1}{11}, \tfrac{2}{11}, \tfrac{3}{11}, \tfrac{4}{11}, \tfrac{5}{11}, \tfrac{6}{11}, \tfrac{7}{11}, \tfrac{8}{11}, \tfrac{9}{11}, \tfrac{10}{11}, 1,$$

$$\tfrac{12}{11}, \tfrac{13}{11}, \tfrac{14}{11}, \tfrac{15}{11}, \tfrac{16}{11}, \tfrac{17}{11}, \tfrac{18}{11}, \tfrac{19}{11}, \tfrac{20}{11}, \tfrac{21}{11}, 2\}.$$

The knots are marked on the graph on the left in Fig. 5.6.

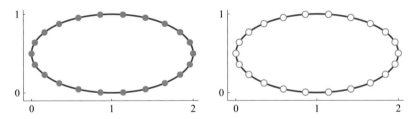

Fig. 5.6 Left: the knot locations on ∂S. Right: the collocation points.

The 22 knots determine 22 functions $b_{i,j}$ for each component of $\tilde{\varphi}$, generating a total of 44 basis functions. The graphs of the $b_{i,j}$ in parametric form are shown in Fig. 5.7.

To compute $\tilde{\varphi}$, we must choose an appropriate number of collocation points, which can be placed anywhere because D, P, φ, and x are smooth on ∂S. The set of values of the parameter t at our selection is

$$\{0, \tfrac{1}{11}, \tfrac{2}{11}, \tfrac{3}{11}, \tfrac{4}{11}, \tfrac{5}{11}, \tfrac{6}{11}, \tfrac{7}{11}, \tfrac{8}{11}, \tfrac{9}{11}, \tfrac{10}{11},$$

$$1, \tfrac{12}{11}, \tfrac{13}{11}, \tfrac{14}{11}, \tfrac{15}{11}, \tfrac{16}{11}, \tfrac{17}{11}, \tfrac{18}{11}, \tfrac{19}{11}, \tfrac{20}{11}, \tfrac{21}{11}\}.$$

These values are marked on the graph on the right in Fig. 5.6.

The same 22 points are used for both components of $\tilde{\varphi}$, yielding a system of 44 constraining equations that enables us to compute the coefficients $c_{\alpha,i,j}$ in (5.8). The 44×44 matrix of this system has a condition number of 24.

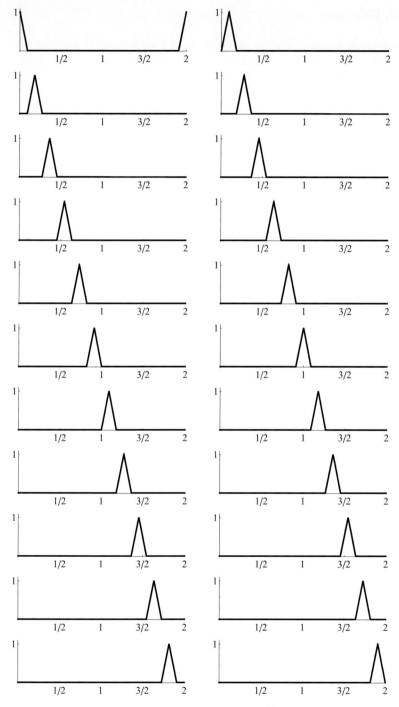

Fig. 5.7 The 22 B-spline basis functions $b_{i,j}[t]$, $0 \le t \le 2$.

5.2.4 Solution

Approximate density. The two components of $\tilde{\varphi}$ are graphed in Figs. 5.8 and 5.9. The latter also shows the spline knot locations. It can be seen that $\tilde{\varphi}$ agrees very well with the exact density φ in Fig. 5.4.

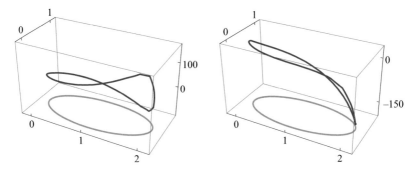

Fig. 5.8 The components of $\tilde{\varphi}[x]$ (Cartesian coordinates).

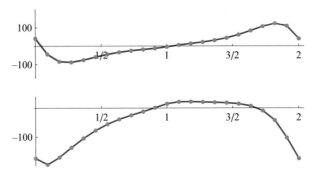

Fig. 5.9 The components of $\tilde{\varphi}[x[t]]$ (parametric form) and the knot locations.

Approximate solution. We use $\tilde{\varphi}$ and $\mathscr{P}$ in (5.7) to compute an approximation $\tilde{u}$ to the exact solution u in S. Its two components are graphed in Fig. 5.10.

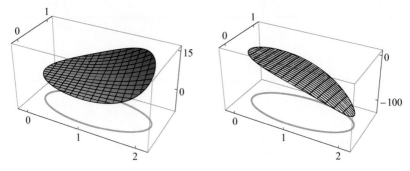

Fig. 5.10 The components of $\tilde{u}[x]$, $x \in S$.

Error analysis. To visualize the difference between the approximate and exact densities, in Fig. 5.11 we graphed $\tilde{\varphi} - \varphi$ relative to the maximum absolute value of φ on ∂S.

The same was done in parametric form in Fig. 5.12, which indicates that the relative error in the two components of $\tilde{\varphi}$ is only about 1%, except possibly at the extreme right of the boundary, where it can exceed 3%.

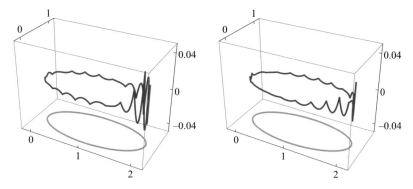

Fig. 5.11 The components of the relative error in $\tilde{\varphi}[x]$ (Cartesian coordinates).

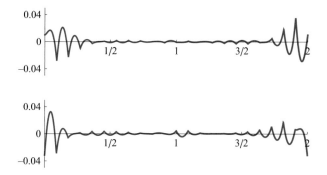

Fig. 5.12 The components of the relative error in $\tilde{\varphi}[x[t]]$ (parametric form).

Figure 5.11 displays a larger error at the extreme left and right segments of the boundary, where ∂S has the largest curvature and where, therefore, the integrand is changing most rapidly. Also, Fig. 5.11 is constructed from the error relative to the maximum absolute value of φ on ∂S, which creates distortion and leads to a larger relative error at the right side of the boundary, where φ attains its maximum absolute value.

5.3 Remarks. (i) The B-spline basis functions had to be modified to produce a basis with periodic continuity at the knot locations $t = 0, 2$. The first function in Fig. 5.7 illustrates the required modification.

(ii) We notice that the equally spaced parametric knot locations and collocation points are, to all intents and purposes, also equally spaced geometrically over the boundary (see Fig. 5.6). Consequently, the error is expected to be essentially the same along the entire boundary contour (see Fig. 5.11). In general, the position of the knots and collocation points should be determined by the boundary ∂S rather than by its parameterization.

(iii) The set of coefficients $\{c_{1,i,j}, c_{2,i,j}\}$ in (5.8), computed with the B-spline basis shown in Fig. 5.7, is

$$\{\{40.1, -44.5, -83.8, -87.3, -75.4, -59.4, -44.8, -33.4,$$
$$- 25.1, -18.9, -12.9, -4.39, 5.41, 13.4, 21.2, 30.8, 43.8, 61.1,$$
$$82.9, 106., 121., 109.\}, \{-173., -194., -170., -136.,$$
$$0.953, 14.5, 21.3, 21.7, 20.6, 19.4, 17.8, -104., -77.4, -55.8,$$
$$- 38.8, -25., -12.4, 14.5, 7.08, -9.24, -42.3, -102.\}\}$$

5.3 Dirichlet Problem in an Ellipse: Piecewise Constant Spline

5.3.1 Summary

The difference between this example and the preceding one lies in the nature of the functions that form the B-spline basis. Using a piecewise constant spline, we construct a family of basis functions with discontinuities at the knot locations and comment on how to make an appropriate choice of (equally spaced) collocation points.

5.3.2 Problem Statement

Domain boundary. ∂S is the ellipse parameterized by

$$\begin{aligned} x1[t] &= 1 + \text{Cos}[\pi t], \\ x2[t] &= \tfrac{1}{2}\text{Sin}[\pi t] + \tfrac{1}{2}, \end{aligned} \qquad 0 \le t \le 2. \qquad (5.9)$$

Its graph can be seen in Fig. 5.18.

Governing equations. The displacement u is the solution of the boundary value problem

$$\begin{aligned} Z_x \diamond u[x] &= 0, \quad x \in S, \\ u[x] &= \mathscr{P}[x], \quad x \in \partial S. \end{aligned}$$

Test solution. The two components of the test solution u given by (5.3) are graphed in Fig. 5.13.

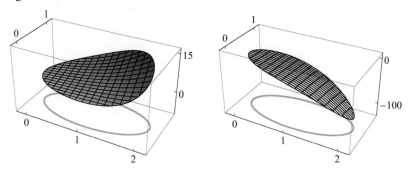

Fig. 5.13 The components of $u[x]$, $x \in S$.

Boundary data function. The function $\mathscr{P}$ is computed as

$$\mathscr{P}[x] = u[x], \quad x \in \partial S.$$

The graphs of its two components are displayed in Figs. 5.14 and 5.15, in Cartesian coordinates and parametric form, respectively.

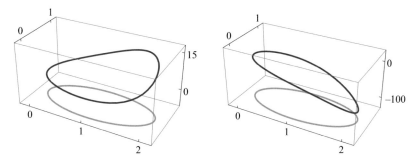

Fig. 5.14 The components of $\mathscr{P}[x]$ (Cartesian coordinates).

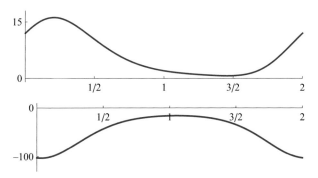

Fig. 5.15 The components of $\mathscr{P}[x[t]]$ (parametric form).

5.3.3 Solution Procedure

Method. We apply the direct method, which reduces the problem to the boundary integral equation

$$V_0(\varphi) = \left(W_0 + \tfrac{1}{2}I\right)\mathscr{P} \quad \text{on } \partial S,$$

coded as

$$\oint_{\Gamma_{\text{Weak}[x]}} D[x,y] \circ \varphi[y]\,d\Gamma_y = \oint_{\Gamma_{\text{CPV}[x]}} P[x,y] \circ \mathscr{P}[y]\,d\Gamma_y + \tfrac{1}{2}\mathscr{P}[x]. \tag{5.10}$$

This equation is solved numerically to obtain an approximate density $\tilde{\varphi}$. In turn, $\tilde{\varphi}$ is used in the representation

$$u[x] = \oint_{\Gamma} D[x,y] \circ \varphi[y]\,d\Gamma_y - \oint_{\Gamma} P[x,y] \circ \mathscr{P}[y]\,d\Gamma_y, \quad x \in S, \tag{5.11}$$

to generate an approximate solution $\tilde{u}$ in S.

Density function. The exact density φ is the boundary stress vector Tu computed from the test solution u and given by (5.4); its components are graphed in Figs. 5.16 and 5.17.

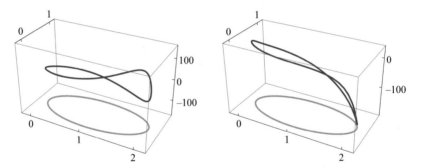

Fig. 5.16 The components of $\varphi[x]$ (Cartesian coordinates).

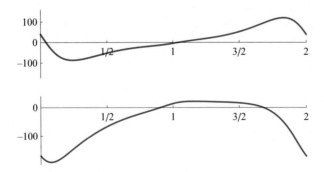

Fig. 5.17 The components of $\varphi[x[t]]$ (parametric form).

Numerical approximation. We compute $\tilde{\varphi}$ by the collocation method with a B-spline basis of elements $b_{i,j}$. Then the approximate density is sought in the form

$$\tilde{\varphi}[x[t]] = \begin{pmatrix} \sum_i \sum_j c_{1,i,j} b_{i,j}[t] \\ \sum_i \sum_j c_{2,i,j} b_{i,j}[t] \end{pmatrix}, \tag{5.12}$$

where the numerical coefficients $c_{\alpha,i,j}$ are determined by substituting (5.12) in (5.10).

We use a piecewise constant spline for which, in terms of the parameterization (5.9), the knot locations and the corresponding smoothness conditions are specified as the set

$$\left\{0, \frac{1}{7}, \frac{2}{7}, \frac{3}{7}, \frac{4}{7}, \frac{5}{7}, \frac{6}{7}, 1, \frac{8}{7}, \frac{9}{7}, \frac{10}{7}, \frac{11}{7}, \frac{12}{7}, \frac{13}{7}, 2\right\}.$$

The knots are marked on the graph on the left in Fig. 5.18.

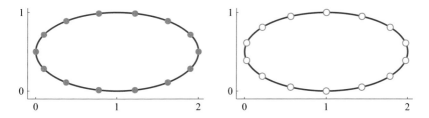

Fig. 5.18 Left: the knot locations on ∂S. Right: the collocation points.

The 14 knots determine 14 functions $b_{i,j}$ for each component of $\tilde{\varphi}$, generating a total of 28 basis functions. The graphs of the $b_{i,j}$ in parametric form are shown in Fig. 5.19.

To compute $\tilde{\varphi}$, we must choose an appropriate number of collocation points. Since D, P, φ, and x are smooth on ∂S, such points can be placed anywhere except at the knots, where the $b_{i,j}$ are discontinuous. Our choice is to position them midway between the knots, which means that the 14 values of t selected as collocation points are the elements of the set

$$\left\{\frac{1}{14}, \frac{3}{14}, \frac{5}{14}, \frac{1}{2}, \frac{9}{14}, \frac{11}{14}, \frac{13}{14}, \frac{15}{14}, \frac{17}{14}, \frac{19}{14}, \frac{3}{2}, \frac{23}{14}, \frac{25}{14}, \frac{27}{14}\right\}.$$

These values are marked on the graph on the right in Fig. 5.18.

The same 14 points are used for both components of the approximate density $\tilde{\varphi}$, yielding a system of 28 constraining equations that enables us to compute the numerical coefficients $c_{\alpha,i,j}$ in (5.12). The 28×28 matrix of this system has a condition number of 11.

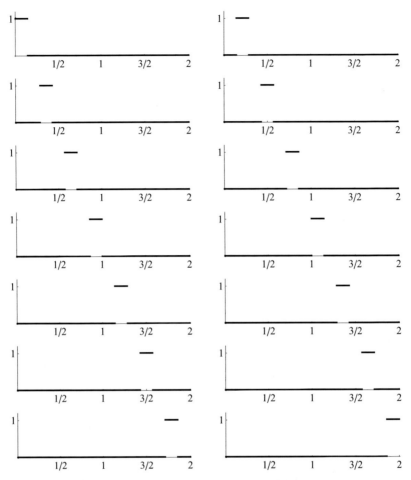

Fig. 5.19 The 14 B-spline basis functions $b_{i,j}[t]$, $0 \le t \le 2$.

5.3.4 Solution

Approximate density. The components of $\tilde{\varphi}$ are graphed in Figs. 5.20 and 5.21.

Refined numerical approximation. From Fig. 5.20 we see that the relative error in the components of $\tilde{\varphi}$ exceeds 20%, which means that we need a finer spline knot spacing. To achieve this, we consider the new knot locations

$$\Big\{0, \frac{1}{25}, \frac{2}{25}, \frac{3}{25}, \frac{4}{25}, \frac{1}{5}, \frac{6}{25}, \frac{7}{25}, \frac{8}{25}, \frac{9}{25}, \frac{2}{5}, \frac{11}{25}, \frac{12}{25}, \frac{13}{25}, \frac{14}{25}, \frac{3}{5}, \frac{16}{25},$$

$$\frac{17}{25}, \frac{18}{25}, \frac{19}{25}, \frac{4}{5}, \frac{21}{25}, \frac{22}{25}, \frac{23}{25}, \frac{24}{25}, 1, \frac{26}{25}, \frac{27}{25}, \frac{28}{25}, \frac{29}{25}, \frac{6}{5}, \frac{31}{25}, \frac{32}{25}, \frac{33}{25},$$

$$\frac{34}{25}, \frac{7}{5}, \frac{36}{25}, \frac{37}{25}, \frac{38}{25}, \frac{39}{25}, \frac{8}{5}, \frac{41}{25}, \frac{42}{25}, \frac{43}{25}, \frac{44}{25}, \frac{9}{5}, \frac{46}{25}, \frac{47}{25}, \frac{48}{25}, \frac{49}{25}, 2\Big\}.$$

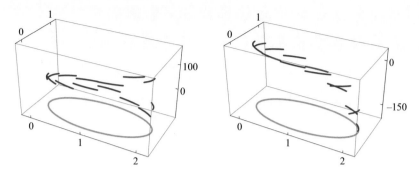

Fig. 5.20 The components of $\tilde{\varphi}[x]$ (Cartesian coordinates).

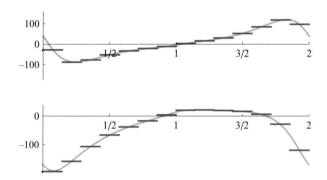

Fig. 5.21 The components of $\tilde{\varphi}[x[t]]$ (parametric form).

These knots are marked on the graph on the left in Fig. 5.22.

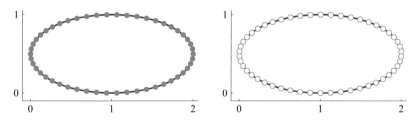

Fig. 5.22 Left: the knot locations on ∂S. Right: the collocation points.

Our chosen 50 collocation points necessary to compute $\tilde{\varphi}$ are marked on the graph on the right in Fig. 5.22. The corresponding values of the parameter t that generate them are the elements of the set

$$\left\{ \frac{1}{50}, \frac{3}{50}, \frac{1}{10}, \frac{7}{50}, \frac{9}{50}, \frac{11}{50}, \frac{13}{50}, \frac{3}{10}, \frac{17}{50}, \frac{19}{50}, \frac{21}{50}, \frac{23}{50}, \frac{1}{2}, \frac{27}{50}, \frac{29}{50}, \frac{31}{50}, \frac{33}{50}, \right.$$

$$\frac{7}{10}, \frac{37}{50}, \frac{39}{50}, \frac{41}{50}, \frac{43}{50}, \frac{9}{10}, \frac{47}{50}, \frac{49}{50}, \frac{51}{50}, \frac{53}{50}, \frac{11}{10}, \frac{57}{50}, \frac{59}{50}, \frac{61}{50}, \frac{63}{50}, \frac{13}{10}, \frac{67}{50},$$

$$\left. \frac{69}{50}, \frac{71}{50}, \frac{73}{50}, \frac{3}{2}, \frac{77}{50}, \frac{79}{50}, \frac{81}{50}, \frac{83}{50}, \frac{17}{10}, \frac{87}{50}, \frac{89}{50}, \frac{91}{50}, \frac{93}{50}, \frac{19}{10}, \frac{97}{50}, \frac{99}{50} \right\}.$$

Using the same 50 collocation points for both components of $\tilde{\varphi}$, we arrive at a system of 100 constraining equations for the coefficients $c_{\alpha,i,j}$ in (5.12). The 100×100 matrix of this system has a condition number of 44.

Refined approximate density. Figure 5.23 shows the graphs of the two components of $\tilde{\varphi}$ computed with the new setup.

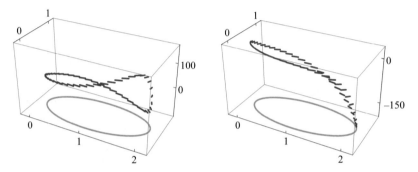

Fig. 5.23 The components of $\tilde{\varphi}[x]$ (Cartesian coordinates).

The components of $\tilde{\varphi}$ and φ are displayed together in Fig. 5.24. These graphs show good agreement between the two functions at the collocation points, which were placed midway between the new spline knots.

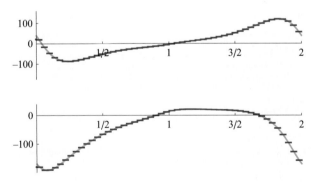

Fig. 5.24 The components of $\tilde{\varphi}[x[t]]$ and $\varphi[x[t]]$ (parametric form).

Approximate solution. We use the refined version of $\tilde{\varphi}$ and $\mathscr{P}$ in (5.11) to compute an approximation $\tilde{u}$ to the exact solution u in S. The graphs of the two components of $\tilde{u}$ are shown in Fig. 5.25.

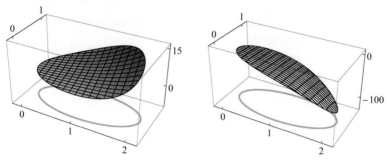

Fig. 5.25 The components of $\tilde{u}[x]$, $x \in S$.

Error analysis. To visualize the difference between the approximate and exact densities, in Fig. 5.26 we graphed $\tilde{\varphi} - \varphi$ relative to the maximum absolute value of φ on ∂S. The same was done in parametric form in Fig. 5.27, which indicates that the relative error in the two components of $\tilde{\varphi}$ is only about 5%, except possibly at the extreme left and right sides of the boundary, where it can exceed that value.

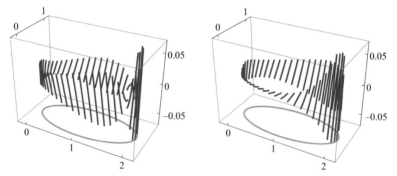

Fig. 5.26 The components of the relative error in $\tilde{\varphi}[x]$ (Cartesian coordinates).

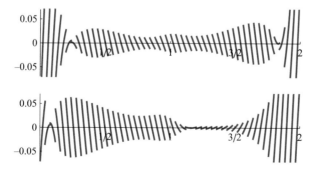

Fig. 5.27 The components of the relative error in $\tilde{\varphi}[x[t]]$ (parametric form).

5.4 Remarks. (i) We notice that the equally spaced parametric knot locations and collocation points are, to all intents and purposes, also equally spaced geometrically over the boundary (see Fig. 5.18). Consequently, the error is expected to be essentially the same along the entire boundary contour (see Fig. 5.23). In general, the position of the knots and collocation points should be determined by the boundary ∂S rather than by its parameterization.

(ii) The set of coefficients $\{c_{1,i,j}, c_{2,i,j}\}$ in (5.12), computed with the B-spline basis shown in Fig. 5.19, is

$$\{\{-27.5, -87.7, -76.6, -51.9, -32.8, -21.,$$
$$-11.2, 3.31, 16.1, 30.1, 51.9, 84.7, 118., 97.1\},$$
$$\{-194., -158., -106., -66., -37.8, -16.9, 4.19,$$
$$20.5, 21.5, 19.6, 16.5, 6.26, -28.3, -119.\}\}.$$

The corresponding set for the B-spline basis constructed with the refined selection of knots and collocation points displayed in Fig. 5.22 is

$$\{\{19.8, -17.2, -47.5, -68.5, -80.9, -86.4, -86.9,$$
$$-83.9, -78.8, -72.4, -65.5, 58.6, -52., -45.8, -40.3,$$
$$-35.4, -31.2, -27.5, -24.3, -21.4, -18.8, -16.2,$$
$$-13.4, -10.2, -6.4, -2.14, 2.15, 6.18, 9.85, 13.2,$$
$$16.6, 20.2, 24., 28.3, 33.1, 38.7, 45., 52.2, 60.3, 69.2,$$
$$78.9, 89., 99.1, 108., 115., 119., 118., 108., 88.5, 57.5\},$$
$$\{-180., -191., -190., -182., -170., -156., -141.,$$
$$-126., -112., -99.5, -87.4, -76.3, -66.3, -57.3,$$
$$-49.2, -41.8, -35.2, -29., -23.3, -17.7, -12.1,$$
$$-6.44, -0.45, 5.7, 11.5, 16.3, 19.5, 21.2, 21.7, 21.6,$$
$$21.3, 20.8, 20.3, 19.7, 19.1, 18.4, 17.5, 16.3, 14.5, 12., 8.36,$$
$$3.2, -3.99, -13.9, -27.3, -45.1, -68.1, -96.3, -127., -158.\}\}.$$

5.4 Dirichlet Problem in an Asymmetric Domain: Piecewise Linear Spline

5.4.1 Summary

This case is similar to the one discussed in Sect. 5.2, with the difference that here the direct method is applied to the Dirichlet problem in an asymmetric domain with a smooth boundary.

5.4.2 Problem Statement

Domain boundary. ∂S is the curve parameterized by

$$\begin{aligned}
x1[t] &= \tfrac{1}{10}\left(7 + 8\mathrm{Cos}[\pi t] + 4\mathrm{Cos}[2\pi t]\right), \\
x2[t] &= \tfrac{1}{50}\left(25 + 2\mathrm{Cos}[3\pi t] + 20\mathrm{Sin}[\pi t]\right),
\end{aligned} \qquad 0 \le t \le 2. \qquad (5.13)$$

Its graph can be seen in Fig. 5.33.

Governing equations. The displacement u is the solution of the boundary value problem

$$\begin{aligned}
Z_x \diamond u[x] &= 0, & x \in S, \\
u[x] &= \mathscr{P}[x], & x \in \partial S.
\end{aligned}$$

Test solution. The two components of the test solution u given by (5.3) are graphed in Fig. 5.28.

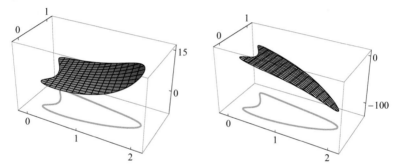

Fig. 5.28 The components of $u[x]$, $x \in S$.

Boundary data function. The function $\mathscr{P}$ is computed from the test solution u as

$$\mathscr{P}[x] = u[x], \qquad x \in \partial S.$$

The graphs of its two components are displayed in Figs. 5.29 and 5.30, in Cartesian coordinates and parametric form, respectively.

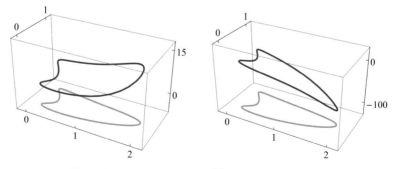

Fig. 5.29 The components of $\mathscr{P}[x]$ (Cartesian coordinates).

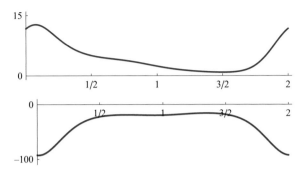

Fig. 5.30 The components of $\mathscr{P}[x[t]]$ (parametric form).

5.4.3 Solution Procedure

Method. We apply the direct method, which reduces the problem to the boundary integral equation

$$V_0(\varphi) = \left(W_0 + \tfrac{1}{2}I\right)\mathscr{P} \quad \text{on } \partial S,$$

coded as

$$\oint_{\Gamma_{\text{Weak}[x]}} D[x,y] \circ \varphi[y]\, d\Gamma_y = \oint_{\Gamma_{\text{CPV}[x]}} P[x,y] \circ \mathscr{P}[y]\, d\Gamma_y + \tfrac{1}{2}\mathscr{P}[x]. \tag{5.14}$$

This equation is solved numerically to obtain an approximate density $\tilde{\varphi}$. In turn, $\tilde{\varphi}$ is used in the representation

$$u[x] = \oint_{\Gamma} D[x,y] \circ \varphi[y]\, d\Gamma_y - \oint_{\Gamma} P[x,y] \circ \mathscr{P}[y]\, d\Gamma_y, \quad x \in S \tag{5.15}$$

to generate an approximate solution $\tilde{u}[x]$, $x \in S$.

Density function. The exact density φ is the boundary stress vector Tu computed from u and given by (5.4); its components are graphed in Figs. 5.31 and 5.32.

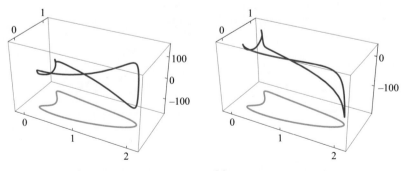

Fig. 5.31 The components of $\varphi[x]$ (Cartesian coordinates).

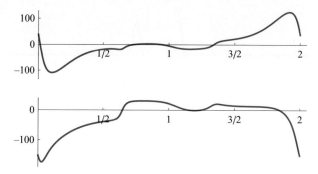

Fig. 5.32 The components of $\varphi[x[t]]$ (parametric form).

Numerical approximation. We compute $\tilde\varphi$ by the collocation method with a B-spline basis of elements $b_{i,j}$. Then the approximate density is sought in the form

$$\tilde\varphi[x[t]] = \begin{pmatrix} \sum_i\sum_j c_{1,i,j}b_{i,j}[t] \\ \sum_i\sum_j c_{2,i,j}b_{i,j}[t] \end{pmatrix}, \tag{5.16}$$

where the numerical coefficients $c_{\alpha,i,j}$ are determined by substituting (5.16) in (5.14).

Since φ is continuous on the boundary, we choose a piecewise linear spline with continuity at all knot locations. In terms of the parameterization (5.13), the primary and secondary knots and the smoothness at their locations are specified as the set

$$\left\{0, \tfrac{1}{14}, \tfrac{1}{7}, \tfrac{3}{14}, \tfrac{2}{7}, \tfrac{5}{14}, \tfrac{3}{7}, \tfrac{1}{2}, \tfrac{4}{7}, \tfrac{9}{14}, \tfrac{5}{7}, \tfrac{11}{14}, \tfrac{6}{7}, \tfrac{13}{14}, 1,\right.$$
$$\left.\tfrac{15}{14}, \tfrac{8}{7}, \tfrac{17}{14}, \tfrac{9}{7}, \tfrac{19}{14}, \tfrac{10}{7}, \tfrac{3}{2}, \tfrac{11}{7}, \tfrac{23}{14}, \tfrac{12}{7}, \tfrac{25}{14}, \tfrac{13}{7}, \tfrac{27}{14}, 2\right\}$$

The knots are marked on the graph on the left in Fig. 5.33.

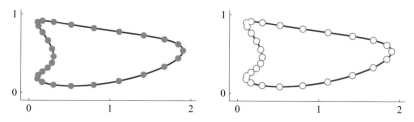

Fig. 5.33 Left: the knot locations on ∂S. Right: the collocation points.

The 28 knots determine 28 functions $b_{i,j}$ for each component of $\tilde\varphi$, generating a total of 56 basis functions. The graphs of the $b_{i,j}$ in parametric form are shown in Fig. 5.34.

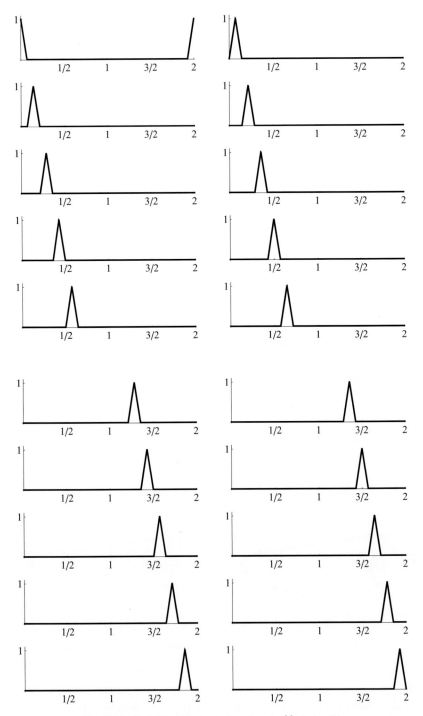

Fig. 5.34 The 28 B-spline basis functions $b_{i,j}[t]$, $0 \le t \le 2$.

To compute $\tilde{\varphi}$, we must choose an appropriate number of collocation points, which can be placed anywhere because D, P, φ, and x are smooth on ∂S. The set of values of the parameter t at our selection is

$$\left\{ 0, \tfrac{1}{14}, \tfrac{1}{7}, \tfrac{3}{14}, \tfrac{2}{7}, \tfrac{5}{14}, \tfrac{3}{7}, \tfrac{1}{2}, \tfrac{4}{7}, \tfrac{9}{14}, \tfrac{5}{7}, \tfrac{11}{14}, \tfrac{6}{7}, \tfrac{13}{14}, 1, \right.$$
$$\left. \tfrac{15}{14}, \tfrac{8}{7}, \tfrac{17}{14}, \tfrac{9}{7}, \tfrac{19}{14}, \tfrac{10}{7}, \tfrac{3}{2}, \tfrac{11}{7}, \tfrac{23}{14}, \tfrac{12}{7}, \tfrac{25}{14}, \tfrac{13}{7}, \tfrac{27}{14} \right\}.$$

These values are marked on the graph on the right in Fig. 5.33.

The same 28 points are used for both components of $\tilde{\varphi}[x[t]]$, yielding a system of 56 constraining equations that enables us to compute the coefficients $c_{\alpha,i,j}$ in (5.16). The 56×56 matrix of this system has a condition number of 74.

5.4.4 Solution

Approximate density. The two components of $\tilde{\varphi}$ are graphed in Figs. 5.35 and 5.36. The latter also shows the spline knot locations. It can be seen that $\tilde{\varphi}$ agrees very well with the exact density φ in Fig. 5.43.

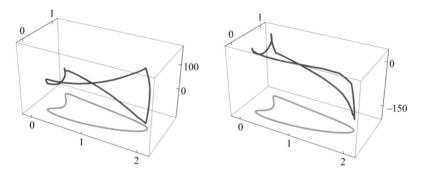

Fig. 5.35 The components of $\tilde{\varphi}[x]$ (Cartesian coordinates).

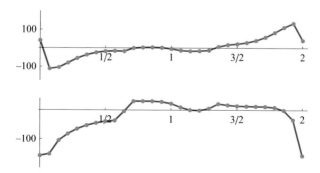

Fig. 5.36 The components of $\tilde{\varphi}[x[t]]$ (parametric form) and the knot locations.

Approximate solution. We use $\tilde{\varphi}$ and $\mathscr{P}$ in (5.15) to compute an approximation $\tilde{u}$ to the exact solution u in S. The graphs of the two components of $\tilde{u}$ are shown in Fig. 5.37.

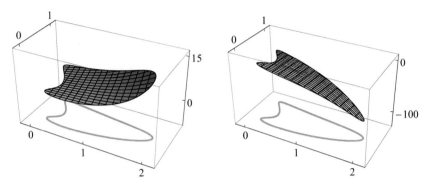

Fig. 5.37 The components of $\tilde{u}[x]$, $x \in S$.

Error analysis. To visualize the difference between the approximate and exact densities, in Fig. 5.38 we graphed $\tilde{\varphi} - \varphi$ relative to the maximum absolute value of φ on ∂S. The same was done in parametric form in Fig. 5.39, which indicates that the relative error in the two components of $\tilde{\varphi}$ is only about 1.5%, except possibly at the extreme left and right sides of the boundary, where it can exceed 5%. This anomaly is explained by the fact that ∂S has the largest curvature on those segments, inducing the fastest variation in the integrand and, therefore, a significant increase in the error. Additionally, Figs. 5.38 and 5.39 are constructed from the error relative to the maximum absolute value of $\varphi]$ on ∂S, which creates distortion and leads to a larger relative error on the extreme right segment of the boundary, where φ attains its maximum absolute value.

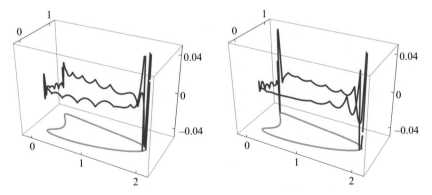

Fig. 5.38 The components of the relative error in $\tilde{\varphi}[x]$ (Cartesian coordinates).

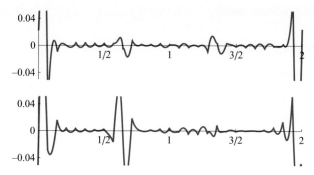

Fig. 5.39 The components of the relative error in $\bar{\varphi}[x[t]]$ (parametric form).

5.5 Remarks. (i) The B-spline basis functions had to be modified to produce a basis with periodic continuity at the knot locations $t = 0, 2$. The first function in Fig. 5.34 illustrates the required modification.

(ii) We notice that the equally spaced parametric knot locations and collocation points are not equally spaced geometrically over ∂S (see Figs. 5.33). Hence, the error is expected to be much larger on the right side of the boundary, where these points are farther apart.

(iii) The set of coefficients $\{c_{1,i,j}, c_{2,i,j}\}$ in (5.16), computed with the B-spline basis shown in Fig. 5.34, is

$$\{\{40.5, -112., -104., -80.1, -55.2, -36.8, -24.9, -18.1, -16.2,$$
$$-17.7, -0.61, 2.42, 3.18, 1.43, -4.91, -13.8, -17.2, -16.2,$$
$$-12.8, 6.89, 17.3, 21.7, 28.9, 40.4, 58.1, 83.1, 111., 133.\},$$
$$\{-157., -152., -105., -82.2, -65., -52.9, -44.7, -39., -36.3,$$
$$-4.42, 30.3, 30.4, 30.3, 28.3, 22.1, 9.55, 0.148, -1.41,$$
$$5.3, 21.3, 17., 14.2, 13.2, 12.8, 12., 9.07, -1.03, -33.6\}\}.$$

5.5 Dirichlet Problem in an Ellipse: Computational Difficulties

5.5.1 Summary

The boundary value problem and its analytic setup are the same in this example as in Sect. 5.2. The difference is that here we investigate the effect of both extreme and minor ill-conditioning, and discuss potential procedures for mitigating it.

5.5.2 Problem Statement

Domain boundary. The curve ∂S is the ellipse parameterized by

$$x1[t] = 1 + \text{Cos}[\pi t], \quad x2[t] = \tfrac{1}{2}\text{Sin}[\pi t] + \tfrac{1}{2}, \quad 0 \le t \le 2. \tag{5.17}$$

Its graph can be seen in Fig. 5.45.

Governing equations. The displacement u is the solution of the boundary value problem

$$Z_x \diamond u[x] = 0, \quad x \in S,$$
$$u[x] = \mathscr{P}[x], \quad x \in \partial S.$$

Test solution. The two components of the test solution u given by (5.3) are graphed in Fig. 5.40.

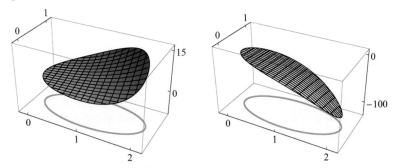

Fig. 5.40 The components of $u[x]$, $x \in S$.

Boundary data function. The function $\mathscr{P}$ is computed from the test solution u as

$$\mathscr{P}[x] = u[x], \quad x \in \partial S.$$

The graphs of its two components are displayed in Figs. 5.41 and 5.42, in Cartesian coordinates and parametric form, respectively.

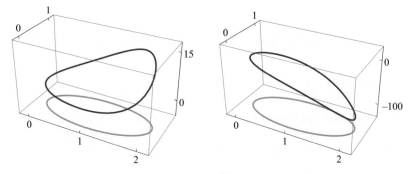

Fig. 5.41 The components of $\mathscr{P}[x]$ (Cartesian coordinates).

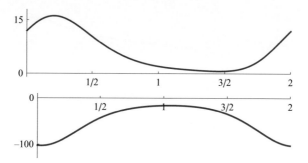

Fig. 5.42 The components of $\mathscr{P}[x[t]]$ (parametric form).

5.5.3 Solution Procedure

Method. We apply the direct method, which reduces the problem to the boundary integral equation

$$V_0(\varphi) = \left(W_0 + \tfrac{1}{2}I\right)\mathscr{P} \quad \text{on } \partial S,$$

coded as

$$\oint_{\Gamma_{\text{Weak}[x]}} D[x,y] \circ \varphi[y]\, d\Gamma_y = \oint_{\Gamma_{\text{CPV}[x]}} P[x,y] \circ \mathscr{P}[y]\, d\Gamma_y + \tfrac{1}{2}\,\mathscr{P}[x]. \tag{5.18}$$

This equation is solved numerically to obtain an approximate density $\tilde{\varphi}$. In turn, $\tilde{\varphi}$ is used in the representation

$$u[x] = \oint_{\Gamma} D[x,y] \circ \varphi[y]\, d\Gamma_y - \oint_{\Gamma} P[x,y] \circ \mathscr{P}[y]\, d\Gamma_y, \quad x \in S \tag{5.19}$$

to generate an approximate solution $\tilde{u}$ in S.

Density function. The exact density φ is the boundary stress vector Tu computed from the test solution u and given by (5.4); its components are graphed in Figs. 5.43 and 5.44.

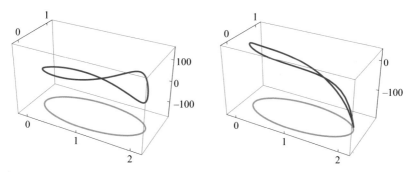

Fig. 5.43 The components of $\varphi[x]$ (Cartesian coordinates).

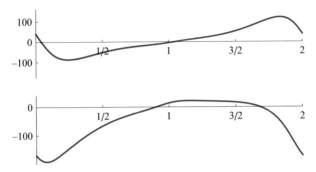

Fig. 5.44 The components of $\varphi[x[t]]$ (parametric form).

Numerical approximation. We compute $\tilde{\varphi}$ by the collocation method with a
B-spline basis of elements $b_{i,j}$. Then the approximate density is sought in the form

$$\tilde{\varphi}[x[t]] = \begin{pmatrix} \sum_i \sum_j c_{1,i,j} b_{i,j}[t] \\ \sum_i \sum_j c_{2,i,j} b_{i,j}[t] \end{pmatrix}, \tag{5.20}$$

where the numerical coefficients $c_{\alpha,i,j}$ are determined by substituting (5.20) in
(5.18).

Since φ is continuous on the boundary, we use a piecewise linear spline with
continuity at all knot locations. In terms of the parameterization (5.17), the knots
and the smoothness at their locations are specified as the set

$$\left\{0, \tfrac{1}{8}, \tfrac{1}{4}, \tfrac{3}{8}, \tfrac{1}{2}, \tfrac{5}{8}, \tfrac{3}{4}, \tfrac{7}{8}, 1, \tfrac{9}{8}, \tfrac{5}{4}, \tfrac{11}{8}, \tfrac{3}{2}, \tfrac{13}{8}, \tfrac{7}{4}, \tfrac{15}{8}, 2\right\}.$$

The knots are marked on the graph on the left in Fig. 5.45.

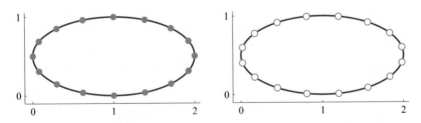

Fig. 5.45 Left: the knot locations on ∂S. Right: the collocation points.

The 16 knots determine 16 functions $b_{i,j}$ for each component of $\tilde{\varphi}$, generating a
total of 32 basis functions. The graphs of the $b_{i,j}$ in parametric form are shown in
Fig. 5.46.

To compute $\tilde{\varphi}$, we must choose an appropriate number of collocation points,
which can be placed anywhere because D, P, φ, and x are smooth on ∂S. The set of

16 values of the parameter t at our selection is

$$\left\{ \frac{1}{16},\ \frac{3}{16},\ \frac{5}{16},\ \frac{7}{16},\ \frac{9}{16},\ \frac{11}{16},\ \frac{13}{16},\ \frac{15}{16},\ \frac{17}{16},\ \frac{19}{16},\ \frac{21}{16},\ \frac{23}{16},\ \frac{25}{16},\ \frac{27}{16},\ \frac{29}{16},\ \frac{31}{16}\ \right\}.$$

These values are marked on the graph on the right in Fig. 5.45.

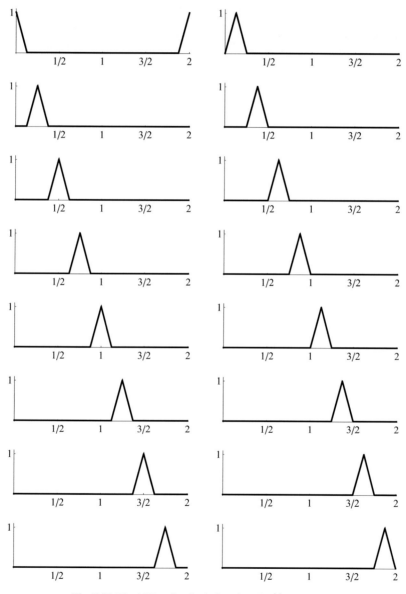

Fig. 5.46 The 16 B-spline basis functions $b_{i,j}[t]$, $0 \le t \le 2$.

Conveniently, the collocation points are placed midway between the spline knots, where the $b_{i,j}$ are not just continuous, but analytic. We use the same 16 points for both components of $\tilde{\varphi}$, yielding a system of 32 constraining equations that enables us to compute the coefficients $c_{\alpha,i,j}$ in (5.20). The 32×32 matrix of this system has a condition number of 28,198.

5.5.4 Solution

Approximate density. The two components of $\tilde{\varphi}$ are graphed in parametric form in Fig. 5.47, where the spline knot locations are also shown.

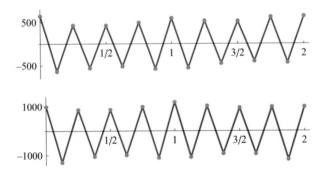

Fig. 5.47 The components of $\tilde{\varphi}[x[t]]$ (parametric form) and the knot locations.

Ill-conditioning. The graphs in Fig. 5.47 bear no resemblance to those of the exact density φ displayed in Fig. 5.44. The first indication that something was wrong was the high condition number 28,198 of the coefficient matrix, which predicted extreme ill-conditioning. Also, the graphs themselves show classic signs of ill-conditioning. The singular value decomposition of the coefficient matrix contains two values that are essentially zero. As a consequence, the rank of the matrix is 2 units less than full rank, so the matrix is singular. The problem is caused by a combination of extreme symmetry in the domain, basis functions, and collocation points. This symmetry is broken if we move the collocation points away from the midpoint between the spline knots.

We select new collocation points by shifting the previous ones by a very small amount (in this case, $1/1000$) away from their original positions. Obviously, the new set

$$\left\{ \frac{127}{2000}, \frac{377}{2000}, \frac{627}{2000}, \frac{877}{2000}, \frac{1127}{2000}, \frac{1377}{2000}, \frac{1627}{2000}, \frac{1877}{2000}, \right.$$

$$\left. \frac{2127}{2000}, \frac{2377}{2000}, \frac{2627}{2000}, \frac{2877}{2000}, \frac{3127}{2000}, \frac{3377}{2000}, \frac{3627}{2000}, \frac{3877}{2000} \right\}$$

looks to the naked eye essentially the same as the original one on the right in Fig. 5.45, but the modified numerical problem has a coefficient matrix with a condition number of 673. Although the components of the new approximate density $\tilde{\varphi}$, graphed in Fig. 5.48, still show signs of ill-conditioning, the true solution is beginning to emerge.

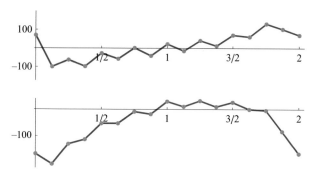

Fig. 5.48 The components of the new $\tilde{\varphi}[x[t]]$, (parametric form) and the knot locations.

In Sect. 5.2, the collocation points for this problem were chosen to coincide with the spline knots, as opposed to the midpoint locations. That choice generated a coefficient matrix with a condition number of 24, meaning that the ill-conditioning had been removed and that the corresponding solution was very accurate.

In Sect. 5.3, the same problem was solved with piecewise constant splines. There, the collocation points were positioned midway between the spline knot locations, which resulted in good conditioning and a good outcome. For that example, choosing the collocation points to be the spline knots would produce the same ill-conditioning as exhibited in the piecewise linear spline version with midpoint locations. The appropriate choice for piecewise constant splines is exactly the opposite of the appropriate choice for piecewise linear splines. This shows that there are no simple rules applicable in all circumstances.

When ill-conditioning is suspected, another approach to deal with the situation is to create an overdetermined system by adding more collocation points. To illustrate this, we double the number of collocation points in the previous case from 16 to 32, choosing the new 32 points to correspond to the set of t values

$$\left\{ \frac{1}{100}, \frac{29}{400}, \frac{27}{200}, \frac{79}{400}, \frac{13}{50}, \frac{129}{400}, \frac{77}{200}, \frac{179}{400}, \frac{51}{100}, \frac{229}{400}, \frac{127}{200}, \frac{279}{400}, \frac{19}{25}, \frac{329}{400}, \frac{177}{200}, \frac{379}{400}, \right.$$

$$\left. \frac{101}{100}, \frac{429}{400}, \frac{227}{200}, \frac{479}{400}, \frac{63}{50}, \frac{529}{400}, \frac{277}{200}, \frac{579}{400}, \frac{151}{100}, \frac{629}{400}, \frac{327}{200}, \frac{679}{400}, \frac{44}{25}, \frac{729}{400}, \frac{377}{200}, \frac{779}{400} \right\}.$$

Keeping the same B-spline basis and knots but using the new set of collocation points yields an overdetermined system with a 64×32 coefficient matrix that has a condition number of 22. The graphs in Fig. 5.49 represent the new approximate density $\tilde{\varphi}$ constructed from the solution of this system with the pseudo-inverse, and include the spline knot locations.

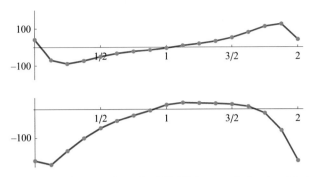

Fig. 5.49 The components of the overdetermined $\tilde{\varphi}[x[t]]$ (parametric form) and the knot locations.

Error analysis. This refined approximation $\tilde{\varphi}$ agrees very well with the exact density φ shown in Fig. 5.48. To visualize the difference between the two, in Fig. 5.50 we graphed $\tilde{\varphi} - \varphi$ relative to the maximum absolute value of φ on ∂S. The same was done in parametric form in Fig. 5.51, which shows that the relative error in the two components of $\tilde{\varphi}$ is only about 1%, except possibly on the extreme right side of ∂S, where the relative error can exceed 5%.

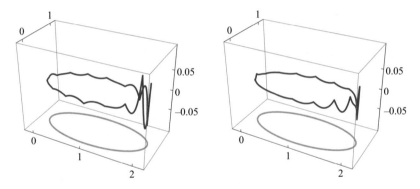

Fig. 5.50 The components of the relative error in $\tilde{\varphi}[x]$ (Cartesian coordinates).

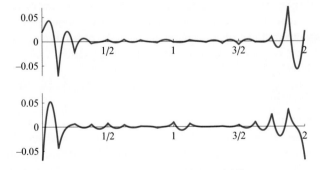

Fig. 5.51 The components of the relative error in $\tilde{\varphi}[x[t]]$ (parametric form).

5.6 Remarks. (i) The aim of the above computations is to show that ill-conditioned problems can arise quite naturally from seemingly appropriate choices. It would be logical, when using piecewise continuous linear spines, to choose the collocation points away from the spline knots, where the basis functions do not have a continuous first-order derivative. But this choice is the very one that creates the ill-conditioning shown here. Also, it is quite possible to make choices that produce only moderate ill-conditioning. This situation is the most difficult to deal with as moderate ill-conditioning is often hard to separate from normal computational inaccuracy. Therefore, it is strongly recommended that careful analysis and vigilance be a part of every numerical problem.

(ii) The set of coefficients $\{c_{1,i,j}, c_{2,i,j}\}$ in (5.20), computed with the B-spline basis shown in Fig. 5.46 for the refined version of $\tilde{\phi}[x[t]]$, is

$$\{\{41.3, -69.9, -89.7, -73.7, -51.5, -34.4, -23.2,$$
$$-15.5, -4.49, 8.82, 19., 32., 51.4, 80., 111., 124.\},$$
$$\{-178., -191., -144., -100., -65.5, -40.5, -21.7,$$
$$-4.51, 15.1, 22.3, 20.9, 19.3, 16.7, 8.91, -13.9, -74.\}\}.$$

5.6 Dirichlet Problem in an Ellipse: Error Evaluation

5.6.1 Summary

We revisit the example in Sect. 5.2 using a piecewise linear spline with equally spaced knots and collocation points, but this time place the emphasis on evaluating the size of the error.

5.6.2 Problem Statement

Domain boundary. The curve ∂S is the ellipse parameterized by

$$x1[t] = 1 + \text{Cos}[\pi t], \quad x2[t] = \tfrac{1}{2}\text{Sin}[\pi t] + \tfrac{1}{2}, \quad 0 \le t \le 2. \tag{5.21}$$

Its graph can be seen in Fig. 5.57.

Governing equations. The displacement u is the solution of the boundary value problem

$$Z_x \diamond u[x] = 0, \quad x \in S,$$
$$u[x] = \mathscr{P}[x], \quad x \in \partial S.$$

Test solution. The two components of the test solution u given by (5.3) are graphed in Fig. 5.52.

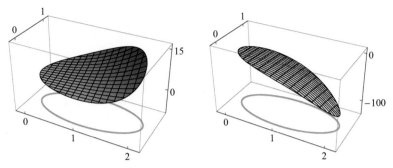

Fig. 5.52 The components of $u[x]$, $x \in S$.

Boundary data function. The function $\mathscr{P}$ is computed from the test solution $u[x]$ as

$$\mathscr{P}[x] = u[x], \quad x \in \partial S.$$

The graphs of its two components are displayed in Figs. 5.53 and 5.54, in Cartesian coordinates and parametric form, respectively.

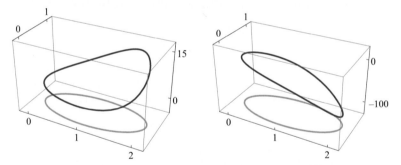

Fig. 5.53 The components of $\mathscr{P}[x]$ (Cartesian coordinates).

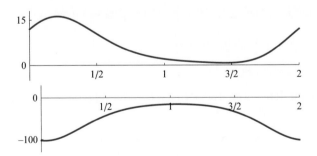

Fig. 5.54 The components of $\mathscr{P}[x[t]]$ (parametric form).

5.6.3 Solution Procedure

Method. We apply the direct method, which reduces the problem to the boundary integral equation

$$V_0(\varphi) = \left(W_0 + \tfrac{1}{2}I\right)\mathscr{P} \quad \text{on } \partial S,$$

coded as

$$\oint_{\Gamma_{\text{Weak}[x]}} D[x,y] \circ \varphi[y] \, d\Gamma_y = \oint_{\Gamma_{\text{CPV}[x]}} P[x,y] \circ \mathscr{P}[y] \, d\Gamma_y + \tfrac{1}{2}\mathscr{P}[x]. \tag{5.22}$$

This equation is solved numerically to obtain an approximate density $\tilde{\varphi}$. In turn, $\tilde{\varphi}$ is used in the representation

$$u[x] = \oint_{\Gamma} D[x,y] \circ \varphi[y] \, d\Gamma_y - \oint_{\Gamma} P[x,y] \circ \mathscr{P}[y] \, d\Gamma_y, \quad x \in S \tag{5.23}$$

to generate an approximate solution $\tilde{u}$ in S.

Density function. The exact density φ is the boundary stress vector Tu computed from the test solution u and given by (5.4); its components are graphed in Figs. 5.55 and 5.56.

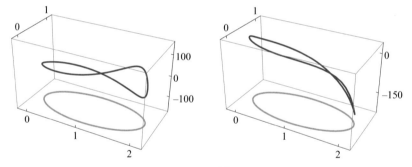

Fig. 5.55 The components of $\varphi[x]$ (Cartesian coordinates).

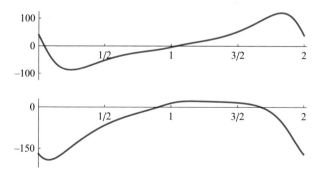

Fig. 5.56 The components of $\varphi[x[t]]$ (parametric form).

Numerical approximation. We compute an approximate density $\tilde{\varphi}$ by the collocation method with a B-spline basis of elements $b_{i,j}$. Then the approximate density is sought in the form

$$\tilde{\varphi}[x[t]] = \begin{pmatrix} \sum_i \sum_j c_{1,i,j} b_{i,j}[t] \\ \sum_i \sum_j c_{2,i,j} b_{i,j}[t] \end{pmatrix}, \tag{5.24}$$

where the numerical coefficients $c_{\alpha,i,j}$ are determined by substituting (5.24) in (5.22).

Since φ is continuous on the boundary, we use a piecewise linear spline with continuity at all knot locations. In terms of the parameterization (5.21), the knots and the smoothness at their locations are specified as the set

$$\left\{ 0, \tfrac{1}{5}, \tfrac{2}{5}, \tfrac{3}{5}, \tfrac{4}{5}, 1, \tfrac{6}{5}, \tfrac{7}{5}, \tfrac{8}{5}, \tfrac{9}{5}, 2 \right\}.$$

The knots are marked on the graph on the left in Fig. 5.57.

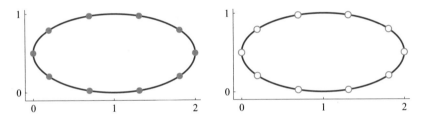

Fig. 5.57 Left: the knot locations on ∂S. Right: the collocation points.

The 10 knots determine 10 functions $b_{i,j}$ for each component of $\tilde{\varphi}$, generating a total of 20 basis functions. The graphs of the $b_{i,j}$ in parametric form are shown in Fig. 5.58.

To compute $\tilde{\varphi}$, we must choose an appropriate number of collocation points, which can be placed anywhere because D, P, φ, and x are smooth on ∂S. We choose these points to coincide with the knots, so the set of values of the parameter t at our selection is

$$\left\{ 0, \tfrac{1}{5}, \tfrac{2}{5}, \tfrac{3}{5}, \tfrac{4}{5}, 1, \tfrac{6}{5}, \tfrac{7}{5}, \tfrac{8}{5}, \tfrac{9}{5} \right\}.$$

These values are marked on the graph on the right in Fig. 5.57.

The same 10 points are used for both components of $\tilde{\varphi}$, yielding a system of 20 constraining equations that enables us to compute the coefficients $c_{\alpha,i,j}$ in (5.24). The 20×20 matrix of this system has a condition number of 9.

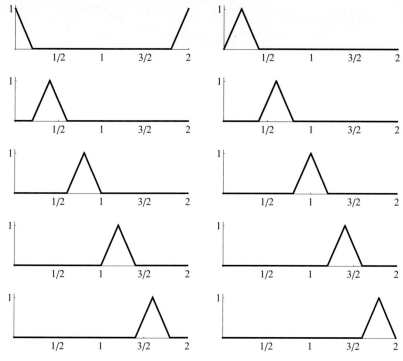

Fig. 5.58 The 10 B-spline basis functions $b_{i,j}[t]$, $0 \le t \le 2$.

5.6.4 Solution

Approximate density. The two components of $\tilde{\varphi}$ are graphed in Figs. 5.59 and 5.60.

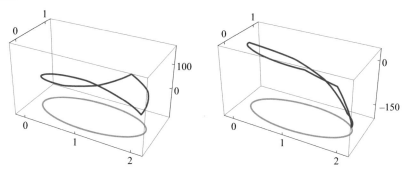

Fig. 5.59 The components of $\tilde{\varphi}[x]$ (Cartesian coordinates).

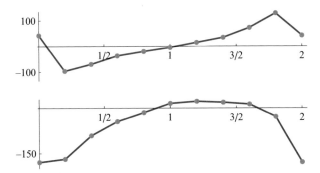

Fig. 5.60 The components of $\tilde{\varphi}[x[t]]$ (parametric form) and the knot locations.

We can now compare $\tilde{\varphi}$ with the exact density φ displayed in Figs. 5.4 and 5.5.

Error analysis. To visualize the difference between the approximate and exact densities, in Fig. 5.61 we graphed the function $\tilde{\varphi} - \varphi$ relative to the maximum absolute value of φ on ∂S. The same was done in parametric form in Fig. 5.62, which indicates that the relative error in the two components of $\tilde{\varphi}$ is only about 5%, except possibly at the extreme right side of the boundary, where it can exceed 15%.

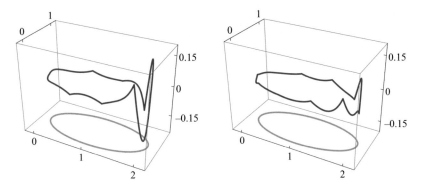

Fig. 5.61 The components of the relative error in $\tilde{\varphi}[x]$ (Cartesian coordinates).

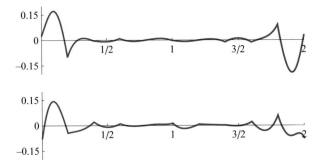

Fig. 5.62 The components of the relative error in $\tilde{\varphi}[x[t]]$ (parametric form).

5.6.5 $O(h^2)$ Analysis

It is well known that the optimal order of approximation for piecewise linear splines used in interpolation is $O(h^2)$, where h is the spacing between adjacent knots. Here, we conduct a computational study to measure the order of convergence in the direct boundary element method for the Dirichlet problem, which reduces to equation $\mathcal{D}_D^+$. The procedure consists in analyzing the rate of improvement of the error as h decreases. Since a global measurement of the error is required, we perform all the error measurements in the L^2-norm. For the example in this section, using our 10 basis functions with $h = 0.2$, we obtain the component error values

$$\|\tilde{\varphi}_1 - \varphi_1\|_2 = 9.74,$$
$$\|\tilde{\varphi}_2 - \varphi_2\|_2 = 6.14.$$

We did a sequence of computational runs with an increasing number of basis functions, namely

$$20, 30, 40, 50, 60, 70, 80, 90, 100,$$

and calculated the L^2-norm error for the two components in each case. The results are displayed in Fig. 5.63.

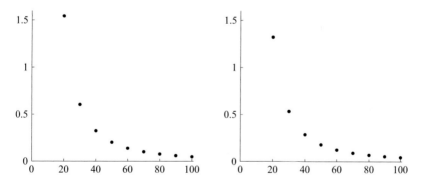

Fig. 5.63 The values of $\|\tilde{\varphi}_\alpha - \varphi_\alpha\|_2$ as functions of basis size.

As expected, the error decreases when the number of basis functions increases, which corresponds to a decrease in the size of h. To see if this decrease is $O(h^2)$, we attempt to fit the model

$$\|\tilde{\varphi} - \varphi\|_2 = Ch^n$$

to the error. Taking logarithms on both sides yields

$$\text{Log}[\|\tilde{\varphi} - \varphi\|_2] = \text{Log}[C] + n\text{Log}[h].$$

Figure 5.64 shows the data points for $\text{Log}[\|\tilde{\varphi}_\alpha - \varphi_\alpha\|_2]$ as functions of $\text{Log}[h]$.

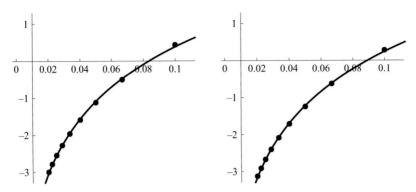

Fig. 5.64 The values of $\text{Log}[\|\tilde{\varphi}_\alpha - \varphi_\alpha\|_2]$ as functions of $\text{Log}[h]$.

The least-squares fit to this data leads to the values

$$\text{Log}[\|\tilde{\varphi}_1 - \varphi_1\|_2] \cong 5.25 + 2.12\text{Log}[h],$$
$$\text{Log}[\|\tilde{\varphi}_2 - \varphi_2\|_2] \cong 5.09 + 2.11\text{Log}[h].$$

When these models, superimposed on the data points in Fig. 5.64, are expressed in traditional form, they produce the approximations

$$\|\tilde{\varphi}_1 - \varphi_1\|_2 \cong 196h^{2.12},$$
$$\|\tilde{\varphi}_2 - \varphi_2\|_2 \cong 166h^{2.11}.$$

Our analysis strongly supports the conjecture that the collocation method with piecewise linear splines is indeed a second-order method.

5.7 Remarks. (i) The above computational experiment has been performed for only one boundary integral method (the direct method), one boundary element method (the collocation method), and one boundary value problem (the Dirichlet problem). Additional work is needed to see if the result can be generalized to other choices of potential applications.

(ii) The set of coefficients $\{c_{1,i,j}, c_{2,i,j}\}$ in (5.24), computed with the the B-spline basis shown in Fig. 5.58, is

$$\{\{42.3, -96.2, -69.4, -36.7, -20., -4.57, 14.9, 34.4, 72.5, 129.\},$$
$$\{-179., -168., -90.8, -44.2, -14.8, 15.5, 22.4, 19., 12.8, -28.7\}\}.$$

5.7 Dirichlet Problem in a Domain with Corners: Piecewise Cubic Spline

5.7.1 Summary

This example illustrates the application of the direct method to a Dirichlet problem in a domain with a continuous and asymmetric boundary that has three corners. The numerical procedure is, in general, the same as in the example discussed in Sect. 5.2, but here the method is implemented by means of a piecewise cubic spline with equally spaced knots and collocation points, and basis functions with discontinuities at the knots.

5.7.2 Problem Statement

Domain boundary. ∂S is the curve parameterized by

$$
x1[t] = \begin{cases} \frac{1}{12}(3+5t+48t^2-32t^3), & 0 \leq t \leq 1, \\ \frac{1}{4}(15-7t), & 1 < t \leq 2,, \\ \frac{1}{4}(5-2t)^2, & 2 < t \leq 3, \end{cases}
$$

$$(5.25)$$

$$
x2[t] = \begin{cases} \frac{5}{8}t, & 0 \leq t \leq 1, \\ \frac{1}{88}(86-127t+128t^2-32t^3), & 1 < t \leq 2, \\ 3-t, & 2 < t \leq 3. \end{cases}
$$

Its graph can be seen in Fig. 5.70.

Governing equations. The displacement u is the solution of the boundary value problem

$$Z_x \diamond u[x] = 0, \quad x \in S,$$
$$u[x] = \mathscr{P}[x], \quad x \in \partial S.$$

Test solution. The two components of the test solution u given by (5.3) are graphed in Fig. 5.65.

Boundary data function. The function $\mathscr{P}$ is computed from the test solution u as

$$\mathscr{P}[x] = u[x], \quad x \in \partial S.$$

The graphs of its two components are displayed in Figs. 5.66 and 5.67, in Cartesian coordinates and parametric form, respectively.

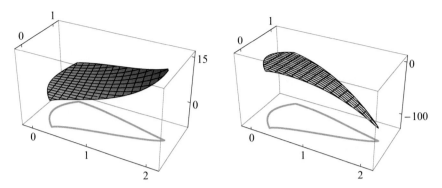

Fig. 5.65 The components of $u[x]$, $x \in S$.

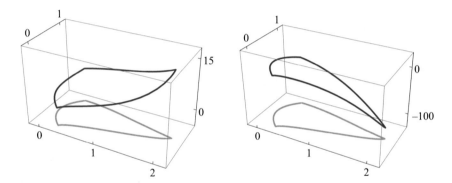

Fig. 5.66 The components of $\mathscr{P}[x]$ (Cartesian coordinates).

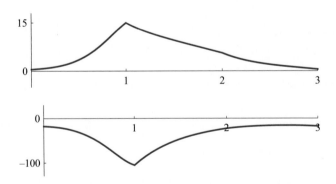

Fig. 5.67 The components of $\mathscr{P}[x[t]]$ (parametric form).

5.7.3 Solution Procedure

Method. We apply the direct method, which reduces the problem to the boundary integral equation

$$V_0(\varphi) = \left(W_0 + \tfrac{1}{2}I\right)\mathscr{P} \quad \text{on } \partial S,$$

coded as

$$\oint_{\Gamma_{\text{Weak}[x]}} D[x,y] \circ \varphi[y] \, d\Gamma_y = \oint_{\Gamma_{\text{CPV}[x]}} P[x,y] \circ \mathscr{P}[y] \, d\Gamma_y + \tfrac{1}{2}\mathscr{P}[x]. \tag{5.26}$$

This equation is solved numerically to obtain an approximate density $\tilde{\varphi}$. In turn, $\tilde{\varphi}$ is used in the representation

$$u[x] = \oint_{\Gamma} D[x,y] \circ \varphi[y] \, d\Gamma_y - \oint_{\Gamma} P[x,y] \circ \mathscr{P}[y] \, d\Gamma_y, \quad x \in S \tag{5.27}$$

to generate an approximate solution $\tilde{u}$ in S.

Density function. The exact density φ is the boundary stress vector Tu computed from the test solution u and given by (5.4); its components are graphed in Figs. 5.68 and 5.69.

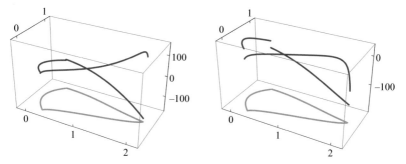

Fig. 5.68 The components of $\varphi[x]$ (Cartesian coordinates).

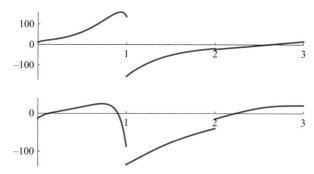

Fig. 5.69 The components of $\varphi[x[t]]$ (parametric form).

Numerical approximation. We compute $\tilde{\varphi}$ by the collocation method with a B-spline basis of elements $b_{i,j}$. Then the approximate density is sought in the form

$$\tilde{\varphi}[x[t]] = \begin{pmatrix} \sum_i \sum_j c_{1,i,j} b_{i,j}[t] \\ \sum_i \sum_j c_{2,i,j} b_{i,j}[t] \end{pmatrix}, \tag{5.28}$$

where the numerical coefficients $c_{\alpha,i,j}$ are determined by substituting (5.28) in (5.26).

Since φ is discontinuous at the three corners of the boundary, the $b_{i,j}$ need to exhibit the same feature. In our case, we choose a piecewise cubic spline that is twice continuously differentiable at the secondary knots and discontinuous at the primary knot locations $t = 0, 1, 2, 3$. In terms of the parameterization (5.25), the primary and secondary knots and the smoothness at their locations are specified as the set

$$\{0, 0, 0, 0\tfrac{1}{2}, 1, 1, 1, 1, \tfrac{3}{2}, 2, 2, 2, 2, \tfrac{5}{2}, 3, 3, 3, 3\}.$$

The knots are marked on the graph on the left in Fig. 5.70.

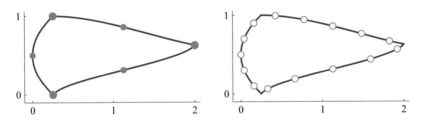

Fig. 5.70 Left: the knot locations on ∂S. Right: the collocation points.

The 6 knots determine 15 functions $b_{i,j}$ for each component of $\tilde{\varphi}$, generating a total of 30 basis functions. The graphs of the $b_{i,j}$ in parametric form are shown in Fig. 5.71.

To compute $\tilde{\varphi}$, we must choose an appropriate number of collocation points, which can be placed anywhere on ∂S except at the corners $t = 0, 1, 2, 3$, where φ is expected to be discontinuous. The set of 15 values of the parameter t at our selection is

$$\left\{ \tfrac{1}{10}, \tfrac{3}{10}, \tfrac{1}{2}, \tfrac{7}{10}, \tfrac{9}{10}, \tfrac{11}{10}, \tfrac{13}{10}, \tfrac{3}{2}, \tfrac{17}{10}, \tfrac{19}{10}, \tfrac{21}{10}, \tfrac{23}{10}, \tfrac{5}{2}, \tfrac{27}{10}, \tfrac{29}{10} \right\}.$$

These values are marked on the graph on the right in Fig. 5.70.

The same 15 points are used for both components of $\tilde{\varphi}$, yielding a system of 30 constraining equations that enables us to compute the coefficients $c_{\alpha,i,j}$ in (5.28). The 30×30 matrix of this system has a condition number of 124.

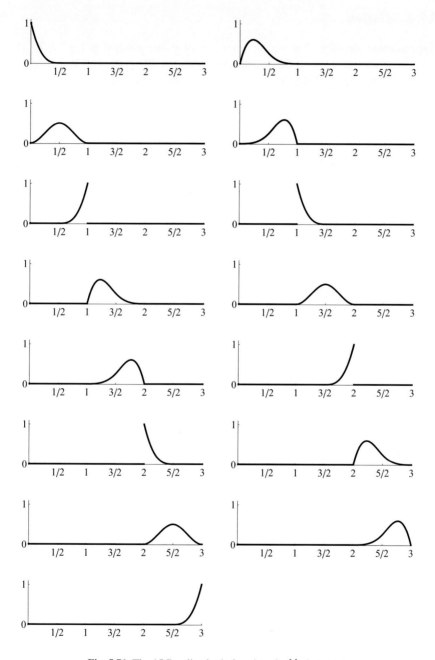

Fig. 5.71 The 15 B-spline basis functions $b_{i,j}[t]$, $0 \leq t \leq 3$.

5.7.4 Solution

Approximate density. The two components of $\tilde{\varphi}$ are graphed in Figs. 5.72 and 5.73. The latter also shows the spline knot locations. It can be seen that $\tilde{\varphi}$ agrees very well with the exact density $\varphi = Tu$ in Figs. 5.68 and 5.69.

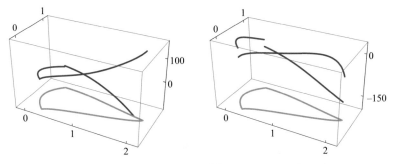

Fig. 5.72 The components of $\tilde{\varphi}[x]$ (Cartesian coordinates).

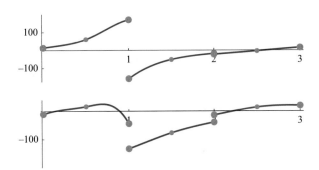

Fig. 5.73 The components of $\tilde{\varphi}[x[t]]$ (parametric form) and the knot locations.

Approximate solution. We use $\tilde{\varphi}$ and $\mathscr{P}$ in (5.27) to compute an approximation $\tilde{u}$ to the exact solution u in S. The graphs of the two components of $\tilde{u}$ are shown in Fig. 5.74.

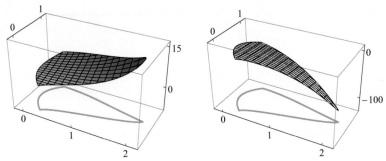

Fig. 5.74 The components of $\tilde{u}[x]$, $x \in S$.

Error analysis. To visualize the difference between the approximate and exact densities, in Fig. 5.75 we graphed $\tilde{\varphi} - \varphi$ relative to the maximum absolute value of φ on ∂S. The same was done in parametric form in Fig. 5.76, which indicates that the relative error in the two components of $\tilde{\varphi}$ is only about 1%, except possibly at the extreme right segment of the boundary, where it can exceed 5%.

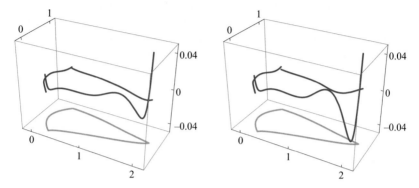

Fig. 5.75 The components of the relative error in $\tilde{\varphi}[x]$ (Cartesian coordinates).

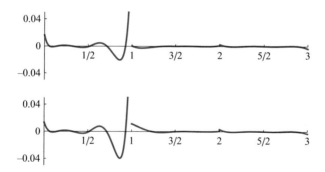

Fig. 5.76 The components of the relative error in $\tilde{\varphi}[x[t]]$ (parametric form).

5.8 Remark. The set of coefficients $\{c_{1,i,j}, c_{2,i,j}\}$ in (5.28), computed with the B-spline basis shown in Fig. 5.71, is

$$\{\{13.4, 25.6, 33.4, 149., 169., -157., -98.6, -41.2,$$
$$-25.5, -18.5, -22.8, -16.6, -4.93, 9.33, 15.\},$$
$$\{-10.9, 6.8, 6.86, 45.8, -44.7, -133., -114.,$$
$$-72.1, -49.6, -39.3, -14., -4.22, 19.5, 21.4, 20.1\}\}.$$

5.8 Dirichlet Problem in a Domain with Corners: Piecewise Quintic Spline

5.8.1 Summary

The analytic setup of the problem in this example is the same as in Sect. 5.7, with the same asymmetric boundary curve ∂S. We show how the accuracy of the method can be improved even when the boundary has corners if we increase the degree of the spline as well as the number of collocation points. We implement the collocation method using a piecewise fifth-degree polynomial spline over equally spaced knots and collocation points, and discuss the accuracy and ill-conditioning of the procedure.

5.8.2 Problem Statement

Domain boundary. ∂S is the curve parameterized by

$$x1[t] = \begin{cases} \frac{1}{12}(3+5t+48t^2-32t^3), & 0 \le t \le 1, \\ \frac{1}{4}(15-7t), & 1 < t \le 2, , \\ \frac{1}{4}(5-2t)^2, & 2 < t \le 3, \end{cases}$$

$$x2[t] = \begin{cases} \frac{5}{8}t, & 0 \le t \le 1, \\ \frac{1}{88}(86-127t+128t^2-32t^3), & 1 < t \le 2, \\ 3-t, & 2 < t \le 3. \end{cases}$$

(5.29)

Its graph can be seen in Fig. 5.82.

Governing equations. The displacement u is the solution of the boundary value problem

$$Z_x \diamond u[x] = 0, \quad x \in S,$$
$$u[x] = \mathscr{P}[x], \quad x \in \partial S.$$

Test solution. The two components of the test solution u given by (5.3) are graphed in Fig. 5.77.

Boundary data function. The function $\mathscr{P}$ is computed from the test solution u as

$$\mathscr{P}[x] = u[x], \quad x \in \partial S.$$

The graphs of its two components are displayed in Figs. 5.78 and 5.79, in Cartesian coordinates and parametric form, respectively.

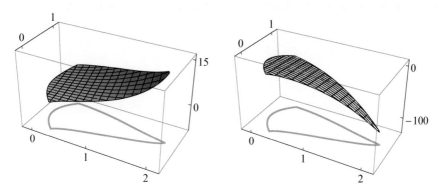

Fig. 5.77 The components of $u[x]$, $x \in S$.

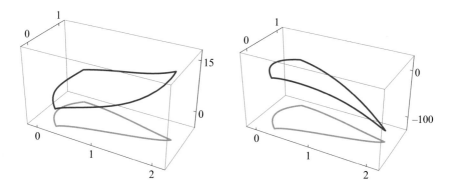

Fig. 5.78 The components of $\mathscr{P}[x]$ (Cartesian coordinates).

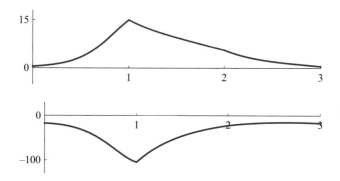

Fig. 5.79 The components of $\mathscr{P}[x[t]]$ (parametric form).

5.8.3 Solution Procedure

Method. We apply the direct method, which reduces the problem to the boundary integral equation

$$V_0(\varphi) = \left(W_0 + \tfrac{1}{2}I\right)\mathscr{P} \quad \text{on } \partial S,$$

coded as

$$\oint_{\Gamma_{\text{Weak}[x]}} D[x,y] \circ \varphi[y]\, d\Gamma_y = \oint_{\Gamma_{\text{CPV}[x]}} P[x,y] \circ \mathscr{P}[y]\, d\Gamma_y + \tfrac{1}{2}\mathscr{P}[x]. \tag{5.30}$$

This equation is solved numerically to obtain an approximate density $\tilde{\varphi}$. In turn, $\tilde{\varphi}$ is used in the representation

$$u[x] = \oint_{\Gamma} D[x,y] \circ \varphi[y]\, d\Gamma_y - \oint_{\Gamma} P[x,y] \circ \mathscr{P}[y]\, d\Gamma_y, \quad x \in S \tag{5.31}$$

to generate an approximate solution $\tilde{u}$ in S.

Density function. The exact density φ is the boundary stress vector Tu computed from the test solution u and given by (5.4); its components are graphed in Figs. 5.80 and 5.81.

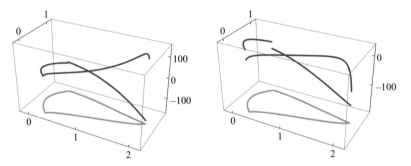

Fig. 5.80 The components of $\varphi[x]$ (Cartesian coordinates).

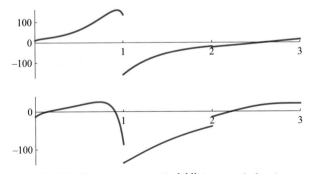

Fig. 5.81 The components of $\varphi[x[t]]$ (parametric form).

Numerical approximation. We compute $\tilde{\varphi}$ by the collocation method with a B-spline basis of elements $b_{i,j}$. Then the approximate density is sought in the form

$$\tilde{\varphi}[x[t]] = \begin{pmatrix} \sum_i \sum_j c_{1,i,j} b_{i,j}[t] \\ \sum_i \sum_j c_{2,i,j} b_{i,j}[t] \end{pmatrix}, \tag{5.32}$$

where the numerical coefficients $c_{\alpha,i,j}$ are determined by substituting (5.32) in (5.30).

Since φ is discontinuous at the three corners of the boundary, the $b_{i,j}$ need to exhibit the same feature. In our case, we choose a piecewise quintic spline that is four times continuously differentiable at the secondary knots and discontinuous at the primary knot locations $t = 0, 1, 2, 3$. In terms of the parameterization (5.29), the primary and secondary knots and the smoothness at their locations are specified as the set

$$\left\{ 0, 0, 0, 0, 0, 0, \tfrac{1}{7}, \tfrac{2}{7}, \tfrac{3}{7}, \tfrac{4}{7}, \tfrac{5}{7}, \tfrac{6}{7}, 1, 1, 1, 1, 1, 1, \tfrac{8}{7}, \tfrac{9}{7}, \tfrac{10}{7}, \right.$$
$$\left. \tfrac{11}{7}, \tfrac{12}{7}, \tfrac{13}{7}, 2, 2, 2, 2, 2, 2, \tfrac{15}{7}, \tfrac{16}{7}, \tfrac{17}{7}, \tfrac{18}{7}, \tfrac{19}{7}, \tfrac{20}{7}, 3, 3, 3, 3, 3, 3 \right\}.$$

The knots are marked on the graph on the left in Fig. 5.82.

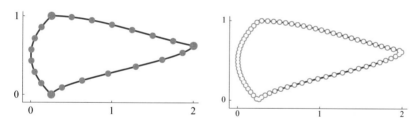

Fig. 5.82 Left: the knot locations on ∂S. Right: the collocation points.

The 22 knots determine 36 functions $b_{i,j}$ for each component of $\tilde{\varphi}$, generating a total of 72 basis functions. The graphs of the $b_{i,j}$ in parametric form are shown in Fig. 5.83.

To compute $\tilde{\varphi}$, we must choose an appropriate number of collocation points, which can be placed anywhere except at the corners $t = 0, 1, 2, 3$, where φ is expected to be discontinuous. The set of 72 values of the parameter t at our selection is

$$\left\{ \tfrac{1}{48}, \tfrac{1}{16}, \tfrac{5}{48}, \tfrac{7}{48}, \tfrac{3}{16}, \tfrac{11}{48}, \tfrac{13}{48}, \tfrac{5}{16}, \tfrac{17}{48}, \tfrac{19}{48}, \tfrac{7}{16}, \tfrac{23}{48}, \tfrac{25}{48}, \tfrac{9}{16}, \tfrac{29}{48}, \tfrac{31}{48}, \tfrac{11}{16}, \tfrac{35}{48}, \tfrac{37}{48}, \right.$$
$$\tfrac{13}{16}, \tfrac{41}{48}, \tfrac{43}{48}, \tfrac{15}{16}, \tfrac{47}{48}, \tfrac{49}{48}, \tfrac{17}{16}, \tfrac{53}{48}, \tfrac{55}{48}, \tfrac{19}{16}, \tfrac{59}{48}, \tfrac{61}{48}, \tfrac{21}{16}, \tfrac{65}{48}, \tfrac{67}{48}, \tfrac{23}{16}, \tfrac{71}{48}, \tfrac{73}{48}, \tfrac{25}{16},$$
$$\tfrac{77}{48}, \tfrac{79}{48}, \tfrac{27}{16}, \tfrac{83}{48}, \tfrac{85}{48}, \tfrac{29}{16}, \tfrac{89}{48}, \tfrac{91}{48}, \tfrac{31}{16}, \tfrac{95}{48}, \tfrac{97}{48}, \tfrac{33}{16}, \tfrac{101}{48}, \tfrac{103}{48}, \tfrac{35}{16}, \tfrac{107}{48}, \tfrac{109}{48}, \tfrac{37}{16},$$
$$\left. \tfrac{113}{48}, \tfrac{115}{48}, \tfrac{39}{16}, \tfrac{119}{48}, \tfrac{121}{48}, \tfrac{41}{16}, \tfrac{125}{48}, \tfrac{127}{48}, \tfrac{43}{16}, \tfrac{131}{48}, \tfrac{133}{48}, \tfrac{45}{16}, \tfrac{137}{48}, \tfrac{139}{48}, \tfrac{47}{16}, \tfrac{143}{48} \right\}.$$

These values are marked on the graph on the right in Fig. 5.82.

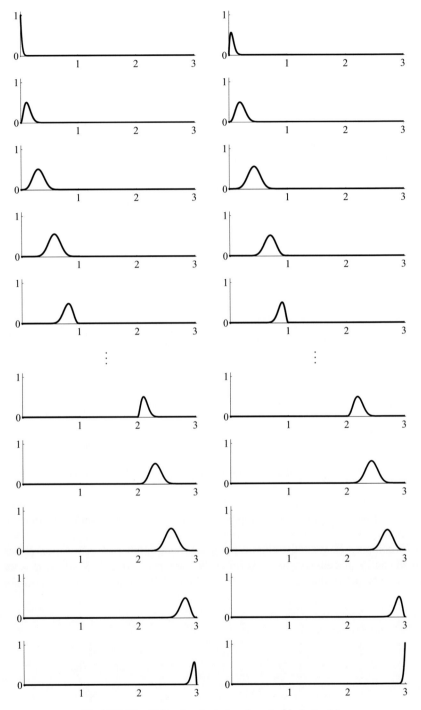

Fig. 5.83 The 36 B-spline basis functions $b_{i,j}[t]$, $0 \le t \le 3$.

The same 72 points are used for both components of $\tilde{\varphi}$, yielding a system of 144 constraining equations that enable us to compute the coefficients $c_{\alpha,i,j}$ in (5.32). The 144×72 matrix of this overdetermined system has a condition number of 712.

5.8.4 Solution

Approximate density. The two components of $\tilde{\varphi}$ are graphed in Figs. 5.84 and 5.85. The latter also shows the spline knot locations. It can be seen that $\tilde{\varphi}$ agrees very well with the exact density $\varphi = Tu$ in Figs. 5.80 and 5.81.

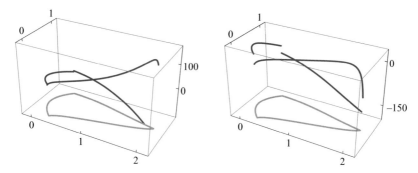

Fig. 5.84 The components of $\tilde{\varphi}[x]$ (Cartesian coordinates).

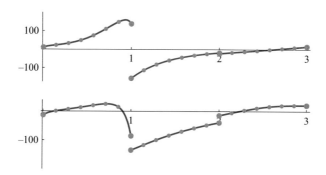

Fig. 5.85 The components of $\tilde{\varphi}[x[t]]$ (parametric form) and the knot locations.

Approximate solution. We use $\tilde{\varphi}$ and $\mathscr{P}$ in (5.31) to compute an approximation $\tilde{u}$ to the exact solution u in S. The graphs of the two components of $\tilde{u}$ are shown in Fig. 5.86. Of course, to the naked eye they look no different than the graphs of the exact solution in Fig.5.74, which were drawn based on computation with a piecewise cubic spline.

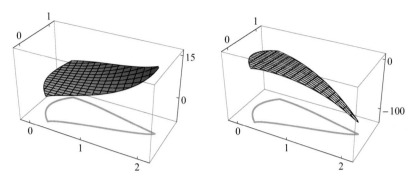

Fig. 5.86 The components of $\tilde{u}[x]$, $x \in S$.

Error analysis. To visualize the difference between the approximate and exact densities, in Fig. 5.87 we graphed $\tilde{\varphi} - \varphi$ relative to the maximum absolute value of φ on ∂S. The same was done in parametric form in Fig. 5.88, which indicates that the relative error in the two components of $\tilde{\varphi}$ is only about 0.005%, except, as expected, at the corners of the boundary.

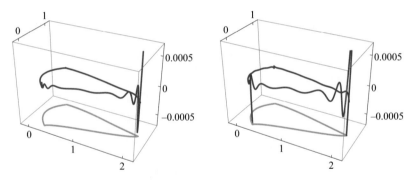

Fig. 5.87 The components of the relative error in $\tilde{\varphi}[x]$ (Cartesian coordinates).

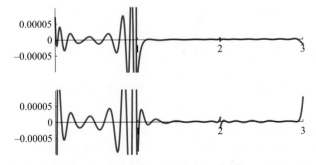

Fig. 5.88 The components of the relative error in $\tilde{\varphi}[x[t]]$ (parametric form).

This high level of accuracy illustrates the benefit of using a piecewise higher-degree polynomial spline for problems with smooth boundary conditions between adjacent corners. The accuracy in the vicinity of the corners is expected to deteriorate significantly. We can partially mitigate this effect by using unequally spaced spline knots and carefully chosen collocation points, as we have done here.

Ill-conditioning. We carried out a second numerical experiment (not shown here) with the collocation points placed midway between the spline knots. This produced a highly ill-conditioned problem. Consequently, we followed the procedure outlined in Sect. 5.5 and augmented the number of collocation points, which reduced from five to three digits the condition number for the coefficient matrix of the system that determines the $c_{\alpha,i,j}$, and thus led to an acceptable computational scheme.

5.9 Remark. The set of the coefficients $\{c_{1,i,j}, c_{2,i,j}\}$ in the approximation $\tilde{\varphi}$ given by (5.32) for the basis functions shown in Fig. 5.83 is

$$\{\{10.8, 15.7, 19.7, 23.8, 31., 46.6, 71.8, 108., 142.,$$
$$163., 160., 136., -157., -146., -128., -104., -80., -58.6,$$
$$-43.3, -32.4, -26., -22.1, -19.8, -18.7, -23.2, -21.9, -19.5,$$
$$-16.3, -12.4, -7.23, -1.43, 4.32, 8.67, 12., 14.3, 15.6\},$$
$$\{-13.2, -7.99, -2.01, 2.54, 6.98, 13.1, 20.3, 27.5, 26.2,$$
$$8.12, -38.4, -87.3, -134., -131., -124., -113., -99.2, -83.5,$$
$$-69.9, -58.2, -50.2, -44.6, -41.1, -39.4, -14.5, -12.5, -8.71,$$
$$-3.28, 3.58, 11.4, 17.2, 20., 20.6, 20.7, 20.6, 20.6\}\}.$$

5.9 Dirichlet Problem in a Square: Piecewise Cubic Spline

5.9.1 Summary

This example illustrates the application of the boundary element method in a case where the boundary conditions are not constructed from a test solution. Consequently, the final result of our computation is not known to us a priori, which raises the question of validation. We aim to answer this question in a 'natural' way. Here, the domain S is a square with Dirichlet conditions prescribed on its boundary. We obtain an approximate solution by means of the direct method and a piecewise cubic spline on equally spaced knots and collocation points. In the next section, we use this solution as input for a Neumann boundary value problem, whose output will enable us to settle the validation issue for the current example.

5.9.2 Problem Statement

Domain boundary. The contour ∂S is the square parameterized by

$$
x1[t] = \begin{cases} t, & 0 \le t \le 1, \\ 1, & 1 < t \le 2, \\ 3-t, & 2 < t \le 3, \\ 0, & 3 < t \le 4, \end{cases} \qquad
x2[t] = \begin{cases} 0, & 0 \le t \le 1, \\ t-1, & 1 < t \le 2, \\ 1, & 2 < t \le 3, \\ 4-t & 3 < t \le 4. \end{cases} \tag{5.33}
$$

Its graph can be seen in Fig. 5.91.

Governing equations. The displacement u is the solution of the boundary value problem

$$
\begin{aligned}
Z_x \diamond u[x] &= 0, \quad x \in S, \\
u[x] &= \mathscr{P}[x], \quad x \in \partial S.
\end{aligned}
$$

Test solution. No test solution is used in this example.

Boundary data function. The function $\mathscr{P}$ is prescribed without reference or connection to any known test solution. Our choice, in parametric form, is

$$
\mathscr{P}_1[x[t]] = \begin{cases} 5 + 10t^2, & 0 \le t \le 1, \\ 20 - 10t + 5t^2, & 1 < t \le 2, \\ -12 + 32t - 8t^2, & 2 < t \le 3, \\ -51 + 42t - 7t^2, & 3 < t \le 4, \end{cases}
$$

$$
\mathscr{P}_2[x[t]] = \begin{cases} 15 - 10t^2, & 0 \le t \le 1, \\ 17 - 24t + 12t^2, & 1 < t \le 2, \\ 37 - 20t + 5t^2, & 2 < t \le 3, \\ -41 + 42t - 7t^2, & 3 < t \le 4. \end{cases}
$$

The graphs of its two components are displayed in Figs. 5.89 and 5.90, in Cartesian coordinates and parametric form, respectively.

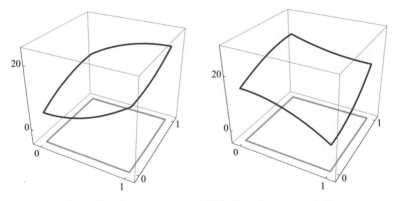

Fig. 5.89 The components of $\mathscr{P}[x]$ (Cartesian coordinates).

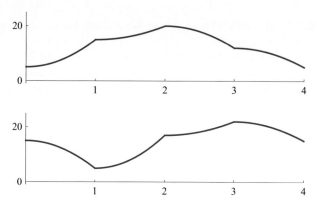

Fig. 5.90 The components of $\mathscr{P}[x[t]]$ (parametric form).

5.9.3 Solution Procedure

Method. We apply the direct method, which reduces the problem to the boundary integral equation

$$V_0(\varphi) = \left(W_0 + \tfrac{1}{2}I\right)\mathscr{P} \quad \text{on } \partial S,$$

coded as

$$\oint_{\Gamma_{\text{Weak}[x]}} D[x,y]\circ\varphi[y]\,d\Gamma_y = \oint_{\Gamma_{\text{CPV}[x]}} P[x,y]\circ\mathscr{P}[y]\,d\Gamma_y + \tfrac{1}{2}\mathscr{P}[x]. \tag{5.34}$$

This equation is solved numerically to obtain an approximate density $\tilde{\varphi}$. In turn, $\tilde{\varphi}$ is used in the representation

$$u[x] = \oint_{\Gamma} D[x,y]\circ\varphi[y]\,d\Gamma_y - \oint_{\Gamma} P[x,y]\circ\mathscr{P}[y]\,d\Gamma_y, \quad x\in S \tag{5.35}$$

to generate an approximate solution $\tilde{u}$ in S.

Density function. This is not known beforehand since we are not using a test solution.

Numerical approximation. We compute $\tilde{\varphi}$ by the collocation method with a B-spline basis of elements $b_{i,j}$. Then the approximate density is sought in the form

$$\tilde{\varphi}[x[t]] = \begin{pmatrix} \sum_i\sum_j c_{1,i,j}b_{i,j}[t] \\ \sum_i\sum_j c_{2,i,j}b_{i,j}[t] \end{pmatrix}, \tag{5.36}$$

where the numerical coefficients $c_{\alpha,i,j}$ are determined by substituting (5.36) in (5.34).

Since φ is expected to be discontinuous at the four corners of ∂S, the b_i, j need to exhibit the same feature. In our case, we choose a piecewise cubic spline that is twice continuously differentiable at the secondary knots and discontinuous at the primary knot locations $t = 0, 1, 2, 3, 4$. In terms of the parameterization (5.33), the primary and secondary knots and smoothness at their locations are specified as the set

$$\left\{0, 0, 0, 0, \tfrac{1}{2}, 1, 1, 1, \tfrac{3}{2}, 2, 2, 2, \tfrac{5}{2}, 3, 3, 3, \tfrac{7}{2}, 4, 4, 4, 4\right\}.$$

The knots are marked on the graph on the left in Fig. 5.91.

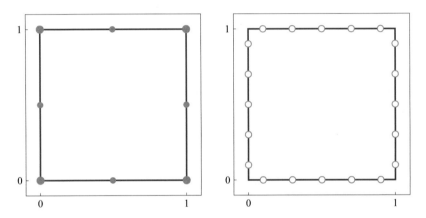

Fig. 5.91 Left: the knot locations on ∂S. Right: the collocation points.

The 9 knots determine 20 functions $b_{i,j}$ for each component of $\tilde{\varphi}$, generating a total of 40 basis functions. The graphs of the $b_{i,j}$ in parametric form are shown in Fig. 5.92.

To compute $\tilde{\varphi}$, we must choose an appropriate number of collocation points, which can be placed anywhere except at the corners $t = 0, 1, 2, 3, 4$, where φ is expected to be discontinuous. The set of 20 values of the parameter t at our selection is

$$\left\{\tfrac{1}{10}, \tfrac{3}{10}, \tfrac{1}{2}, \tfrac{7}{10}, \tfrac{9}{10}, \tfrac{11}{10}, \tfrac{13}{10}, \tfrac{3}{2}, \tfrac{17}{10}, \tfrac{19}{10}, \tfrac{21}{10}, \tfrac{23}{10}, \tfrac{5}{2}, \tfrac{27}{10}, \tfrac{29}{10}, \tfrac{31}{10}, \tfrac{33}{10}, \tfrac{7}{2}, \tfrac{37}{10}, \tfrac{39}{10}\right\}.$$

These values are marked on the graph on the right in Fig. 5.91.

The same 20 points are used for both components of $\tilde{\varphi}$, yielding a system of 40 constraining equations that enables us to compute the coefficients $c_{\alpha,i,j}$ in (5.36). The 40×40 matrix of this system has a condition number of 120.

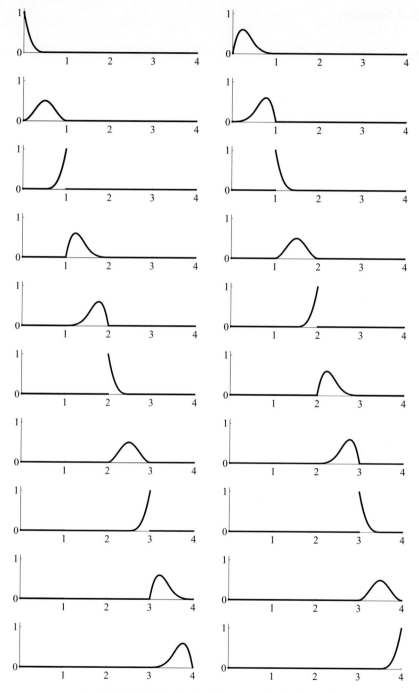

Fig. 5.92 The 20 B-spline basis functions $b_{i,j}[t]$, $0 \le t \le 4$.

5.9.4 Solution

Approximate density. The components of $\tilde{\varphi}$ are graphed in Figs. 5.93 and 5.94.

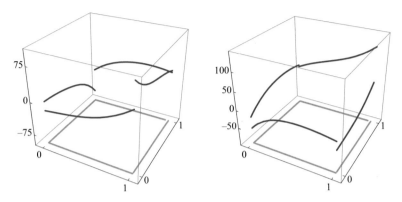

Fig. 5.93 The components of $\tilde{\varphi}[x]$ (Cartesian coordinates).

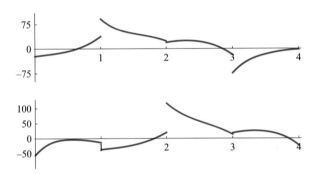

Fig. 5.94 The components of $\tilde{\varphi}[x[t]]$ (parametric form).

Approximate solution. We use $\tilde{\varphi}$ and $\mathscr{P}$ in (5.35) to compute an approximation $\tilde{u}$ to the exact solution u in S. The graphs of the components of $\tilde{u}$ are shown in Fig. 5.95.

Error analysis. The exact solution of this problem is not known; therefore, no comparison can be made between the calculated density $\tilde{\varphi}$ and the exact density φ. However, since φ represents Tu, in Sect. 5.10 we use the approximate density $\tilde{\varphi}$ computed here as the input function to a Neumann boundary value problem; that is,

$$\mathscr{Q}[x[t]] = (\mathrm{Tu})[x[t]] = \tilde{\varphi}[x[t]] \quad \text{on } \partial S.$$

The unknown density ψ in the integral equation for that problem represents the value of the solution u on ∂S; hence, it is the (approximate) data function for the example discussed above, and we can compare it to our choice of function $\mathscr{P}$ for validation purposes.

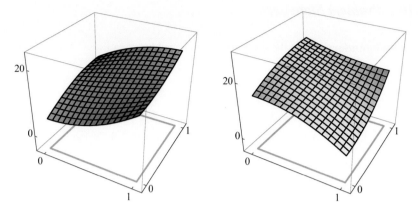

Fig. 5.95 The components of $\tilde{u}[x]$, $x \in S$.

5.10 Remarks. (i) As already mentioned, $\tilde{\varphi}$ is not constructed from a known test so-lution like (5.3). The family of test solutions that we generated in Sect. 4.13, which includes (5.3), exhibits radial symmetry about the point x in polar coordinates, so it is a very limited collection of solutions for the general problem. The paramet-ric functions used to define $\mathscr{P}$ in this example does not have such symmetry and, therefore, it is a good illustration of the general case.

(ii) The set of coefficients $\{c_{1,i,j}, c_{2,i,j}\}$ in (5.36), computed with the B-spline basis shown in Fig. 5.92, is

$$\{\{-22.8, -18.2, -13.3, 12.6, 37.4, 91.3, 61.5, 44.5, 35.5, 23.4,$$
$$18.9, 24.8, 28., 5.41, -18.7, -74.1, -36.9, -13., -3.61, -2.35\},$$
$$\{-55.6, -13., 1.48, -6.75, -13.8, -36.9, -33.4, -26.7, -1.13, 19.6,$$
$$117., 84.7, 54.1, 33.1, 13., 17.8, 25.4, 29.9, 6.16, -23.5\}\}.$$

5.10 Neumann Problem in a Square: Validation Procedure

5.10.1 Summary

This example illustrates the application of the boundary element method to a Neu-mann problem in a square, where the function prescribed on the boundary is the approximate density computed for the Dirichlet problem in Sect. 5.9. The solu-tion is obtained with the direct method by means of a piecewise cubic spline on equally spaced collocation points. The numerical output will be compared to the input Dirichlet data in the preceding example, to help illustrate the validity of our methods when the exact solution is unknown.

5.10.2 Problem Statement

Domain boundary. ∂S is the square parameterized by

$$
x1[t] = \begin{cases} t, & 0 \le t \le 1, \\ 1, & 1 < t \le 2, \\ 3-t, & 2 < t \le 3, \\ 0, & 3 < t \le 4, \end{cases} \qquad x2[t] = \begin{cases} 0, & 0 \le t \le 1, \\ t-1, & 1 < t \le 2, \\ 1, & 2 < t \le 3, \\ 4-t & 3 < t \le 4. \end{cases} \tag{5.37}
$$

Its graph can be seen in Fig. 5.98.

Governing equations. The displacement u is the solution of the boundary value problem

$$
\begin{aligned}
Z_x \diamond u[x] &= 0, \quad x \in S, \\
(\mathrm{T}u)[x] = T_x \diamond u[x] &= \mathscr{Q}[x], \quad x \in \partial S.
\end{aligned}
$$

Test solution. No test solution is used for this example.

Boundary data function. The function $\mathscr{Q}$ is defined by

$$
\mathscr{Q}[x] = \tilde{\varphi}[x], \quad 0 \le t \le 4,
$$

where $\tilde{\varphi}$ is the approximate density computed in Sect. 5.9. The graphs of the two components of $\mathscr{Q}$ are displayed in Figs. 5.96 and 5.97, in Cartesian coordinates and parametric form, respectively.

This choice makes it possible for us to verify the computational accuracy in the boundary element method used to calculate both $\tilde{\varphi}$ and the approximate density $\tilde{\psi}$ in this example. However, no assessment of the exact accuracy can be made because the exact solution is unknown.

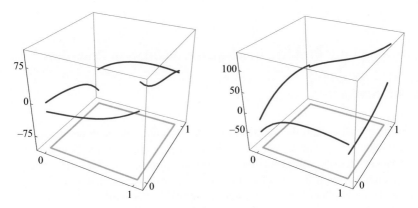

Fig. 5.96 The components of $\mathscr{Q}[x]$ (Cartesian coordinates).

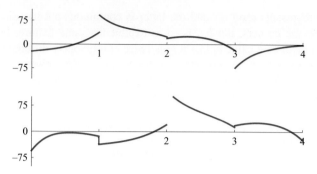

Fig. 5.97 The components of $\mathscr{Q}[x[t]]$ (parametric form).

5.10.3 Solution Procedure

Method. We apply the direct method, which reduces the problem to the boundary integral equation

$$\left(W_0 + \tfrac{1}{2}I\right)\psi - V_0\mathscr{Q} \quad \text{on } \partial S,$$

codes as

$$\oint_{\Gamma_{\mathrm{CPV}[x]}} P[x,y] \circ \psi[y]\,d\Gamma_y + \tfrac{1}{2}\,\psi[x] = \oint_{\Gamma_{\mathrm{Weak}[x]}} D[x,y] \circ \mathscr{Q}[y]\,d\Gamma_y. \tag{5.38}$$

This equation is solved numerically to obtain an approximate density $\tilde{\psi}$. In turn, $\tilde{\psi}$ is used in the representation

$$u[x] = \oint_{\Gamma} D[x,y] \circ \mathscr{Q}[y]\,d\Gamma_y - \oint_{\Gamma} P[x,y] \circ \psi[y]\,d\Gamma_y, \quad x \in S \tag{5.39}$$

to generate an approximate solution $\tilde{u}$ in S.

Density function. This example does not use a test solution u, so the exact density function ψ is not available.

Numerical approximation. We compute $\tilde{\psi}$ by the collocation method with a B-spline basis of elements $b_{i,j}$. Then the approximate density is sought in the form

$$\tilde{\psi}[x[t]] = \begin{pmatrix} \sum_i \sum_j c_{1,i,j} b_{i,j}[t] \\ \sum_i \sum_j c_{2,i,j} b_{i,j}[t] \end{pmatrix}, \tag{5.40}$$

where the numerical coefficients $c_{\alpha,i,j}$ are determined by substituting (5.40) in (5.38).

Since ψ represents u on ∂S and the latter is continuous but has discontinuous derivatives at the corners, the $b_{i,j}$ need to exhibit the same feature. In our case, we choose a piecewise cubic spline that is twice continuously differentiable at the secondary knots and continuous but with discontinuous derivatives at the primary knot locations $t = 0, 1, 2, 3, 4$. In terms of the parameterization (5.37), the knots and the smoothness at their locations are specified as the set

$$\left\{0, 0, 0, \tfrac{1}{2}, 1, 1, 1, \tfrac{3}{2}, 2, 2, 2, \tfrac{5}{2}, 3, 3, 3, \tfrac{7}{2}, 4, 4, 4\right\}.$$

The knots are marked on the graph on the left in Fig. 5.98.

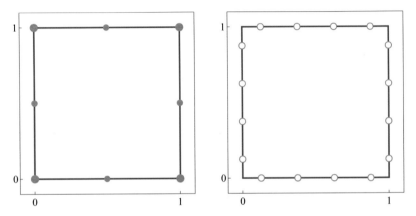

Fig. 5.98 Left: the knot locations on ∂S. Right: the collocation points.

The 9 knots determine 16 functions $b_{i,j}$ for each component of $\tilde{\psi}$, generating a total of 32 basis functions. The graphs of the $b_{i,j}$ in parametric form are shown in Fig. 5.99.

To compute $\tilde{\psi}$, we must choose an appropriate number of collocation points, which can be placed anywhere on ∂S because ψ is expected to be continuous on the boundary. The set of 16 values of the parameter t at our selection is

$$\left\{\tfrac{1}{8}, \tfrac{3}{8}, \tfrac{5}{8}, \tfrac{7}{8}, \tfrac{9}{8}, \tfrac{11}{8}, \tfrac{13}{8}, \tfrac{15}{8}, \tfrac{17}{8}, \tfrac{19}{8}, \tfrac{21}{8}, \tfrac{23}{8}, \tfrac{25}{8}, \tfrac{27}{8}, \tfrac{29}{8}, \tfrac{31}{8}\right\}.$$

These values are marked on the graph on the right in Fig. 5.98.

The same 16 collocation points are used for both components of $\tilde{\psi}$, yielding a system of 32 constraining equations that enable us to compute the coefficients $c_{\alpha,i,j}$ in (5.40). The 32×32 matrix of this system has a condition number of 256.

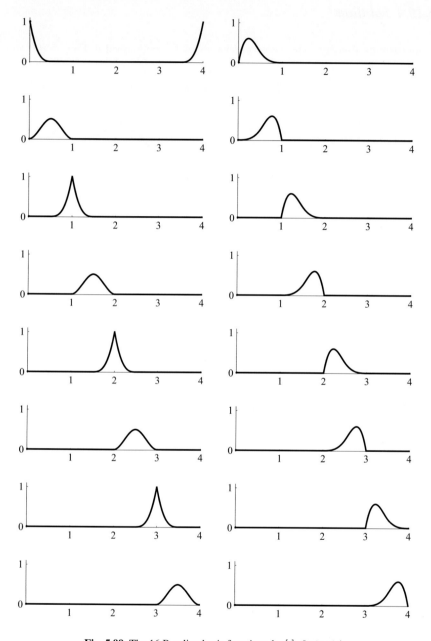

Fig. 5.99 The 16 B-spline basis functions $b_{i,j}[t]$, $0 \leq t \leq 4$.

5.10.4 Solution

Approximate density. The two components of $\tilde{\psi}$ are graphed in Figs. 5.100 and 5.101.

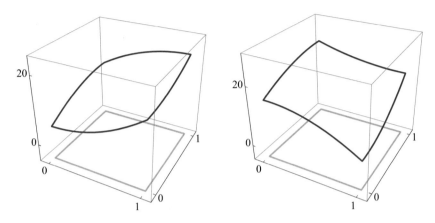

Fig. 5.100 The components of $\tilde{\psi}[x]$ (Cartesian coordinates).

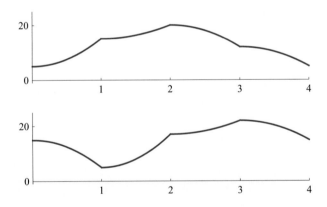

Fig. 5.101 The components of $\tilde{\psi}[x[t]]$ (parametric form).

Approximate solution. We use $\tilde{\psi}$ and $\mathcal{Q}$ in (5.39) to compute an approximation $\tilde{u}$ to the exact solution u in S. The graphs of the two components of $\tilde{u}$ are shown in Fig. 5.102.

Error analysis. In Remark 5.10(i), we mentioned that ψ represents the boundary values of u. Since those values were given by the function $\mathcal{P}$ in the preceding example, and since $\tilde{\psi}$ has been calculated here using an input constructed from $\mathcal{P}$, it is important for us to determine how close both sets of results are.

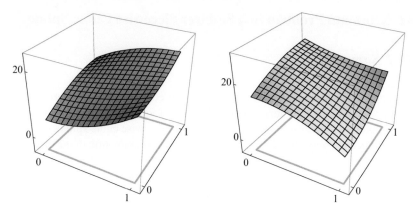

Fig. 5.102 The components of $\tilde{u}[x]$, $x \in S$.

To visualize the difference between $\tilde{\psi}$ and $\mathscr{P}$, in Fig. 5.103 we graphed $\tilde{\psi} - \mathscr{P}$ relative to the maximum absolute value of $\mathscr{P}$ on ∂S. These graphs indicate that the relative error in the two components of $\tilde{\psi}$ is only about 0.2%, except possibly at the corners of the boundary, where it can exceed 0.3%. The fact that the composite of both the Dirichlet and Neumann problems yields such a small error supports the validity of our calculations.

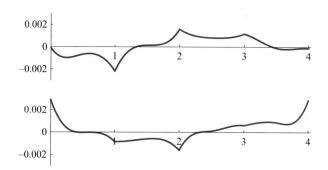

Fig. 5.103 The components of the relative error in $\tilde{\psi}[x[t]]$ (parametric form).

5.11 Remarks. (i) The exact solution of this problem is not known; therefore, no comparison can be made between the calculated density $\tilde{\psi}$ and the exact density ψ.

(ii) The set of coefficients $\{c_{1,i,j}, c_{2,i,j}\}$ in (5.40), computed with the B-spline basis shown in Fig. 5.99, is

$$\{\{5., 5.02, 6.67, 11.6, 15., 15., 15.8, 18.3, 19.9, 19.9, 18.6,$$
$$14.6, 11.9, 11.9, 10.8, 7.33\}, \{14.9, 14.9, 13.3, 8.33, 5.01, 5.01,$$
$$7.01, 13., 17., 16.9, 17.8, 20.3, 21.9, 21.9, 20.8, 17.3\}\}.$$

5.11 Neumann Problem in a Square: Piecewise Cubic Spline

5.11.1 Summary

This example illustrates the application of the direct method to a Neumann problem in a square domain, where the boundary condition is computed from a known test solution. We use the collocation method with piecewise cubic splines on equally spaced knots and collocation points to construct an approximation of the original solution from the computed boundary data function. The results are validated through a comparison between the exact and approximate solutions.

5.11.2 Problem Statement

Domain boundary. The contour ∂S is the square parameterized by

$$
x1[t] = \begin{cases} t, & 0 \le t \le 1, \\ 1, & 1 < t \le 2, \\ 3-t, & 2 < t \le 3, \\ 0, & 3 < t \le 4, \end{cases} \qquad
x2[t] = \begin{cases} 0, & 0 \le t \le 1, \\ t-1, & 1 < t \le 2, \\ 1, & 2 < t \le 3, \\ 4-t & 3 < t \le 4. \end{cases} \tag{5.41}
$$

Its graph can be seen in Fig. 5.109.

Governing equations. The displacement u is the solution of the boundary value problem

$$
Z_x \diamond u[x] = 0, \quad x \in S,
$$
$$
(Tu)[x] = T_x \diamond u[x] = \mathcal{Q}[x], \quad x \in \partial S.
$$

Test solution. The two components of the test solution u given by (5.3) are graphed in Fig. 5.104.

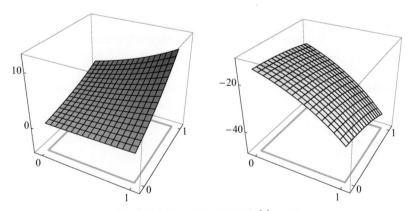

Fig. 5.104 The components of $u[x]$, $x \in S$.

Boundary data function. The function $\mathscr{Q} = \mathrm{T}u$ is given by (5.4). The graphs of its two components are displayed in Figs. 5.105 and 5.106, in Cartesian coordinates and parametric form, respectively.

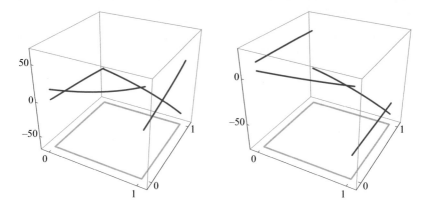

Fig. 5.105 The components of $\mathscr{Q}[x]$ (Cartesian coordinates).

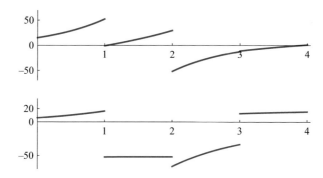

Fig. 5.106 The components of $\mathscr{Q}[x[t]]$ (parametric form).

5.11.3 Solution Procedure

Method. We apply the direct method, which reduces the problem to the boundary integral equation

$$\left(W_0 + \tfrac{1}{2}I\right)\psi - V_0\mathscr{Q} \quad \text{on } \partial S,$$

coded as

$$\oint_{\Gamma_{\text{CPV}[x]}} P[x,y] \circ \psi[y]\, d\Gamma_y + \tfrac{1}{2}\, \psi[x] = \oint_{\Gamma_{\text{Weak}[x]}} D[x,y] \circ \mathscr{Q}[y]\, d\Gamma_y. \tag{5.42}$$

This equation is solved numerically to obtain an approximate density $\tilde{\psi}$. In turn, $\tilde{\psi}$ is used in the representation

$$u[x] = \oint_{\Gamma} D[x,y] \circ \mathscr{Q}[y]\, d\Gamma_y - \oint_{\Gamma} P[x,y] \circ \psi[y]\, d\Gamma_y, \quad x \in S \tag{5.43}$$

to generate an approximate solution $\tilde{u}$ in S.

Density function. The exact density ψ is the restriction to ∂S of the test solution u; its components are graphed in Figs. 5.107 and 5.108.

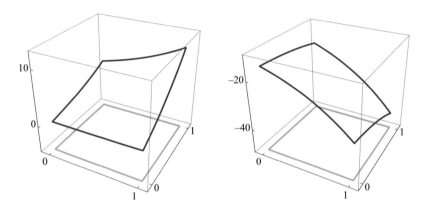

Fig. 5.107 The components of $\psi[x]$ (Cartesian coordinates).

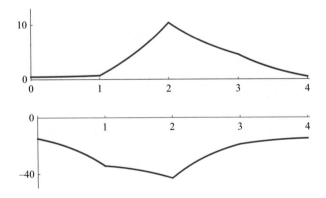

Fig. 5.108 The components of $\psi[x[t]]$ (parametric form).

Numerical approximation. We compute $\tilde{\psi}$ by the collocation method with a B-spline basis of elements $b_{i,j}$. Then the approximate density is sought in the form

$$\tilde{\psi}[x[t]] = \begin{pmatrix} \sum_i \sum_j c_{1,i,j} b_{i,j}[t] \\ \sum_i \sum_j c_{2,i,j} b_{i,j}[t] \end{pmatrix}, \tag{5.44}$$

where the numerical coefficients $c_{\alpha,i,j}$ are determined by substituting (5.44) in (5.42).

Since ψ represents u on ∂S and the latter is continuous but has discontinuous derivatives at the corners, the $b_{i,j}$ need to exhibit the same feature. In our case, we choose a piecewise cubic spline that is twice continuously differentiable at the secondary knots and continuous but with discontinuous derivatives at the primary knot locations $t = 0, 1, 2, 3, 4$. In terms of the parameterization (5.41), the knots and the smoothness at their locations are specified as the set

$$\left\{0, 0, 0, \tfrac{1}{2}, 1, 1, 1, \tfrac{3}{2}, 2, 2, 2, \tfrac{5}{2}, 3, 3, 3, \tfrac{7}{2}, 4, 4, 4\right\}.$$

The knots are marked on the graph on the left in Fig. 5.109.

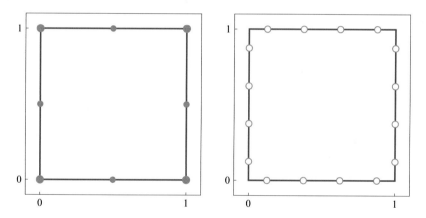

Fig. 5.109 Left: the knot locations on ∂S. Right: the collocation points.

The 8 knots determine 16 functions $b_{i,j}$ for each component of $\tilde{\psi}$, generating a total of 32 basis functions. The graphs of the $b_{i,j}$ in parametric form, modified to enforce the required continuity at the knot locations, are shown in Fig. 5.110.

To compute $\tilde{\psi}$, we must choose an appropriate number of collocation points, which can be placed anywhere because ψ is expected to be continuous on ∂S. We select equally spaced points that avoid the primary knots. The set of 16 values of the parameter t at our selection is

$$\left\{\tfrac{1}{8}, \tfrac{3}{8}, \tfrac{5}{8}, \tfrac{7}{8}, \tfrac{9}{8}, \tfrac{11}{8}, \tfrac{13}{8}, \tfrac{15}{8}, \tfrac{17}{8}, \tfrac{19}{8}, \tfrac{21}{8}, \tfrac{23}{8}, \tfrac{25}{8}, \tfrac{27}{8}, \tfrac{29}{8}, \tfrac{31}{8}\right\}.$$

These values are marked on the graph on the right in Fig. 5.109.

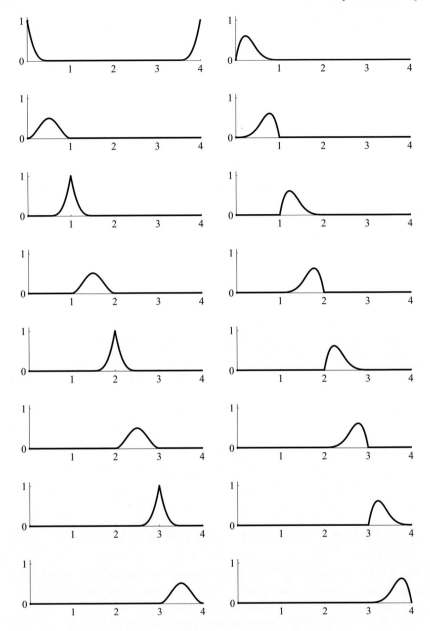

Fig. 5.110 The 16 B-spline basis functions $b_{i,j}[t]$, $0 \leq t \leq 4$.

The same 16 collocation points are used for both components of $\tilde{\psi}$, yielding a system of 32 constraining equations that enable us to compute the coefficients $c_{\alpha,i,j}$ in (5.44). The 32×32 matrix of this system has a condition number of 256.

5.11.4 Solution

Approximate density. The two components of $\tilde{\psi}$ are graphed in Figs. 5.111 and 5.112. The latter also shows the spline knot locations. It can be seen that $\tilde{\psi}$ agrees very well with the exact density ψ in Figs. 5.107 and 5.108.

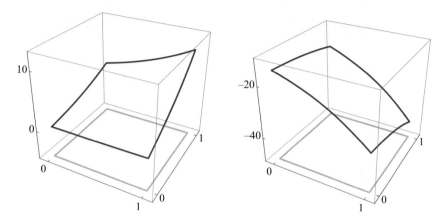

Fig. 5.111 The components of $\tilde{\psi}[x]$ (Cartesian coordinates).

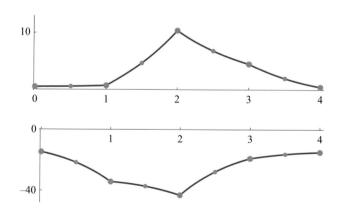

Fig. 5.112 The components of $\tilde{\psi}[x[t]]$ (parametric form).

Approximate solution. We use $\tilde{\psi}$ and $\mathscr{Q}$ in (5.43) to compute an approximation $\tilde{u}$ to the exact solution u in S. The graphs of the two components of $\tilde{u}$ are shown in Fig. 5.113.

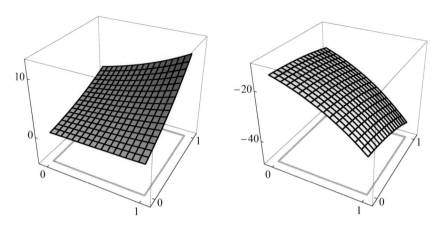

Fig. 5.113 The components of $\tilde{u}[x]$, $x \in S$.

Error analysis. To visualize the difference between the approximate and exact densities, in Fig. 5.114 we graphed $\tilde{\psi} - \psi$ relative to the maximum absolute value of ψ on ∂S.

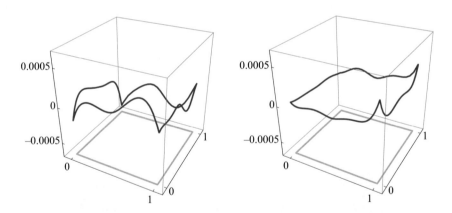

Fig. 5.114 The components of the relative error in $\tilde{\psi}[x]$ (Cartesian coordinates).

The same was done in parametric form in Fig. 5.115, which also shows the primary and secondary knots and indicates that the relative error in the two components of $\tilde{\psi}$ is about 0.05%.

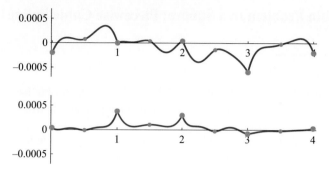

Fig. 5.115 The components of the relative error in $\tilde{\psi}[x[t]]$ (parametric form).

With such a small error on the boundary, we expect an equally good agreement in the interior of the domain. This is indeed confirmed by Fig. 5.116, which displays the graphs of the components of $\tilde{u} - u$ in S.

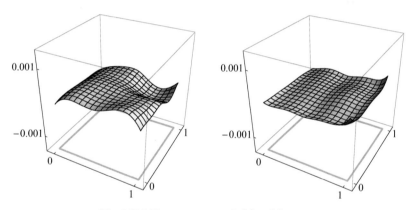

Fig. 5.116 The components of $\tilde{u}[x] - u[x]$, $x \in S$.

5.12 Remark. The set of coefficients $\{c_{1,i,j}, c_{2,i,j}\}$ in (5.44), computed with the B-spline basis shown in Fig. 5.110, is

$$\{\{0.474, 0.493, 0.523, 0.632, 0.706, 1.84, 4.4, 8.07,$$
$$10.3, 8.9, 6.62, 5.15, 4.52, 3.45, 1.82, 0.864\},$$
$$\{-14.8, -16.4, -20.8, -28.8, -34.2, -34.8, -36.7, -40.5,$$
$$-43., -36.4, -26.6, -21.1, -18.9, -17.7, -15.8, -15.\}\}.$$

5.12 Robin Problem in a Square: Piecewise Cubic Spline

5.12.1 Summary

Here, we illustrate the application of the direct method to the Robin boundary value problem in a square domain. We compute the boundary condition from a known test solution, then use the collocation method with a piecewise cubic spline on equally spaced knots and collocation points to approximate the exact solution. The results are validated through a comparison between the approximate and exact solutions.

5.12.2 Problem Statement

Domain boundary. The contour ∂S is the square parameterized by

$$
x1[t] = \begin{cases} t, & 0 \le t \le 1, \\ 1, & 1 < t \le 2, \\ 3-t, & 2 < t \le 3, \\ 0, & 3 < t \le 4, \end{cases} \quad
x2[t] = \begin{cases} 0, & 0 \le t \le 1, \\ t-1, & 1 < t \le 2, \\ 1, & 2 < t \le 3, \\ 4-t & 3 < t \le 4. \end{cases} \tag{5.45}
$$

Its graph can be seen in Fig. 5.122.

Governing equations. The displacement u is the solution of the boundary value problem

$$
Z_x \diamond u[x] = 0, \quad x \in S,
$$
$$
(Tu)[x] + \sigma u[x] = \mathscr{K}[x], \quad x \in \partial S.
$$

Test solution. The two components of the test solution u given by (5.3) are graphed in Fig. 5.117.

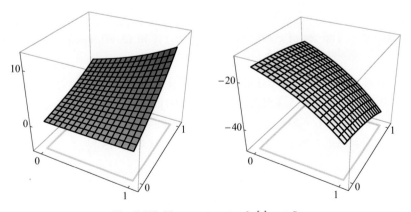

Fig. 5.117 The components of $u[x]$, $x \in S$.

Boundary data function. The function $\mathscr{K}$ is given by

$$(\mathbf{T}u)[x] + \sigma u[x] = \mathscr{K}[x], \quad x \in \partial S,$$

where, for computational purposes, we have chosen the particular 2×2 matrix

$$\sigma = \begin{pmatrix} 2 & 0 \\ 0 & 2 \end{pmatrix}.$$

The graphs of the two components of $\mathscr{K}$ are displayed in Figs. 5.118 and 5.119, in Cartesian coordinates and parametric form, respectively.

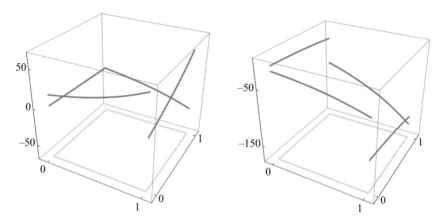

Fig. 5.118 The components of $\mathscr{K}[x]$ (Cartesian coordinates).

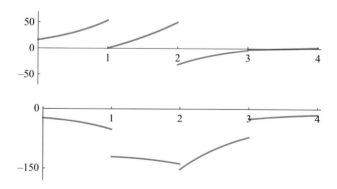

Fig. 5.119 The components of $\mathscr{K}[x[t]]$ (parametric form).

5.12.3 Solution Procedure

Method. We apply the direct method, which reduces the problem to the boundary integral equation

$$\left(W_0 + \tfrac{1}{2} I\right) \psi + V_0(\sigma \psi) = V_0(\mathscr{K}) \quad \text{on } \partial S,$$

coded as

$$\oint_{\Gamma_{\mathrm{CPV}[x]}} P[x,y] \circ \psi[y] \, d\Gamma_y + \oint_{\Gamma_{\mathrm{Weak}[x]}} D[x,y] \circ (\sigma \psi[y]) \, d\Gamma_y + \tfrac{1}{2} \psi[x]$$

$$= \oint_{\Gamma_{\mathrm{Weak}[x]}} D[x,y] \circ \mathscr{K}[y] \, d\Gamma_y. \tag{5.46}$$

This equation is solved numerically to obtain an approximate density $\tilde{\psi}$. In turn, $\tilde{\psi}$ is used in representation

$$u[x] = \oint_{\Gamma} D[x,y] \circ (\mathscr{K} - \sigma \psi[y]) \, d\Gamma_y - \oint_{\Gamma} P[x,y] \circ \psi[y] \, d\Gamma_y, \quad x \in S \tag{5.47}$$

to generate an approximate solution $\tilde{u}$ in S.

Density function. The exact density ψ is the restriction to ∂S of the test solution u; its components are graphed in Figs. 5.120 and 5.121.

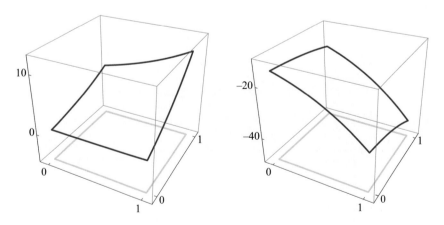

Fig. 5.120 The components of $\psi[x]$, $x \in \partial S$ (Cartesian coordinates).

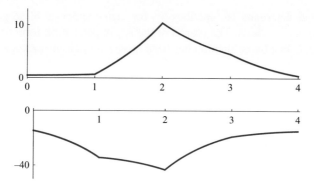

Fig. 5.121 The components of $\psi[x[t]]$, $0 \le t \le 4$ (parametric form).

Numerical approximation. We compute $\tilde{\psi}$ by the collocation method with a B-spline basis of elements $b_{i,j}$. Then the approximate density is sought in the form Then the approximate density is sought in the form

$$\tilde{\psi}[x[t]] = \begin{pmatrix} \sum_i \sum_j c_{1,i,j} b_{i,j}[t] \\ \sum_i \sum_j c_{2,i,j} b_{i,j}[t] \end{pmatrix} , \qquad (5.48)$$

where the coefficients $c_{\alpha,i,j}$ are determined by substituting (5.48) in (5.46).

Since ψ represents u on ∂S and the latter is continuous but has discontinuous derivatives at the corners, the $b_{i,j}$ need to exhibit the same feature. In our case, we choose a piecewise cubic spline that is twice continuously differentiable at the secondary knots and continuous but with discontinuous derivatives at the primary knot locations $t = 0, 1, 2, 3, 4$. In terms of the parameterization (5.45), the knots and the smoothness at their locations are specified as the set

$$\left\{ 0, 0, 0, \tfrac{1}{2}, 1, 1, 1, \tfrac{3}{2}, 2, 2, 2, \tfrac{5}{2}, 3, 3, 3, \tfrac{7}{2}, 4, 4, 4 \right\}.$$

The knots are marked on the graph on the left in Fig. 5.122.

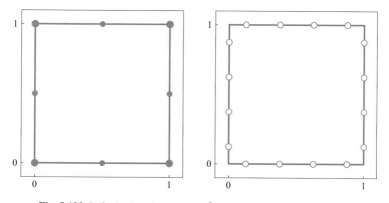

Fig. 5.122 Left: the knot locations on ∂S. Right: the collocation points.

The 8 knots determine 16 functions $b_{i,j}$ for each component of $\tilde{\psi}$, generating a total of 32 basis functions. The graphs of the $b_{i,j}$ in parametric form are shown in Fig. 5.123. The first basis function has been modified to ensure periodic continuity at the knot locations $t = 0, 4$.

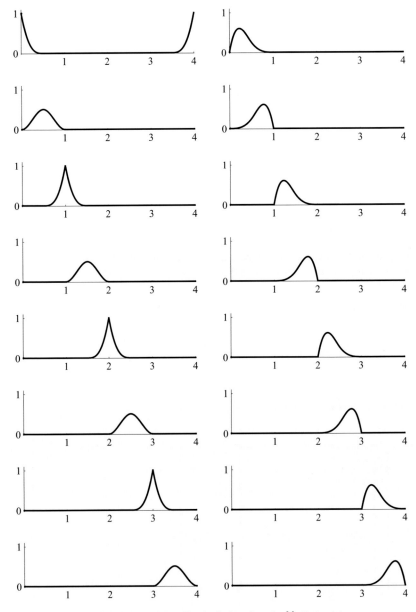

Fig. 5.123 The 16 B-spline basis functions $b_{i,j}[t]$, $0 \leq t \leq 4$.

To compute $\tilde{\psi}$, we must choose an appropriate number of collocation points, which can be placed anywhere because ψ is expected to be continuous on ∂S. We select equally spaced points that avoid the primary knots. The set of 16 values of the parameter t at our selection is

$$\left\{ \tfrac{1}{8}, \tfrac{3}{8}, \tfrac{5}{8}, \tfrac{7}{8}, \tfrac{9}{8}, \tfrac{11}{8}, \tfrac{13}{8}, \tfrac{15}{8}, \tfrac{17}{8}, \tfrac{19}{8}, \tfrac{21}{8}, \tfrac{23}{8}, \tfrac{25}{8}, \tfrac{27}{8}, \tfrac{29}{8}, \tfrac{31}{8} \right\}.$$

These values are marked on the graph on the right in Fig. 5.122.

The same 16 collocation points are used for both components of $\tilde{\psi}$, yielding a system of 32 constraining equations that enables us to compute the coefficients $c_{\alpha,i,j}$ in (5.48). The 32×32 matrix of this system has a condition number of 245.

5.12.4 Solution

Approximate density. The two components of $\tilde{\psi}$ are graphed in Figs. 5.124 and 5.125. The latter also shows the spline knot locations. It can be seen that $\tilde{\psi}$ agrees very well with the exact density $\psi[x] = u[x]$, $x \in \partial S$, shown in Figs. 5.120 and 5.121.

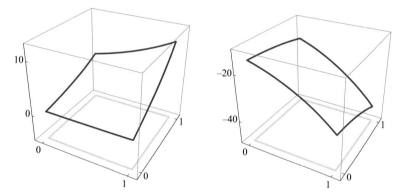

Fig. 5.124 The components of $\tilde{\psi}[x]$ (Cartesian coordinates).

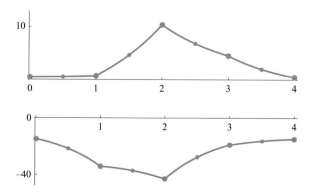

Fig. 5.125 The components of $\tilde{\psi}[x[t]]$ (parametric form).

Approximate solution. We use $\tilde{\psi}$ and $\mathcal{K}$ in (5.48) to compute an approximation $\tilde{u}$ to the exact solution u in S. The graphs of the two components of $\tilde{u}$ are shown in Fig. 5.126.

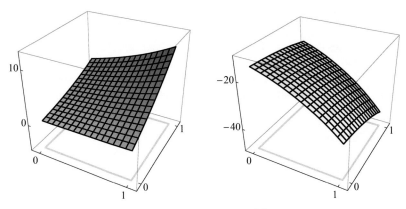

Fig. 5.126 The components of $\tilde{u}[x]$, $x \in S$.

Error analysis. To visualize the difference between the approximate and exact densities, in Fig. 5.127 we graphed $\tilde{\psi} - \psi$ relative to the maximum absolute value of ψ on ∂S.

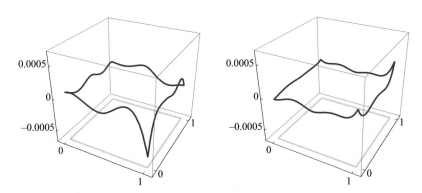

Fig. 5.127 The components of the relative error in $\tilde{\psi}[x]$ (Cartesian coordinates).

The same was done in parametric form in Fig. 5.128, which also shows the primary and secondary knots and indicates that the relative error in the two components of $\tilde{\psi}$ is about 0.05%. With such a small error on the boundary, we expect an equally good agreement in the interior of the domain. This is indeed confirmed by Fig. 5.129, which displays the graphs of the components of $\tilde{u} - u$ in S.

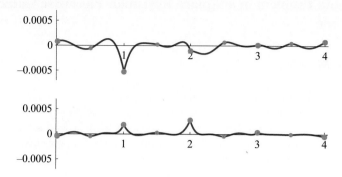

Fig. 5.128 The components of the relative error in $\tilde{\psi}[x[t]]$ (parametric form).

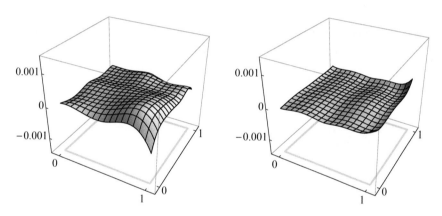

Fig. 5.129 The components of $\tilde{u}[x] - u[x]$, $x \in S$.

5.13 Remark. The set of coefficients $\{c_{1,i,j}, c_{2,i,j}\}$ in (5.48), computed with the B-spline basis shown in Fig. 5.123, is

$$\{\{0.477, 0.491, 0.522, 0.631, 0.701, 1.84, 4.39, 8.07,$$
$$10.3, 8.9, 6.62, 5.15, 4.53, 3.44, 1.83, 0.859\},$$
$$\{-14.8, -16.4, -20.8, -28.8, -34.2, -34.8, -36.7, -40.5,$$
$$-43., -36.4, -26.6, -21.1, -18.9, -17.7, -15.8, -15.\}\}.$$

5.13 Robin Problem in a Square Revisited: Piecewise Cubic Spline

5.13.1 Summary

This example illustrates the application of the alternative direct method to the Robin boundary value problem in a square domain. We compute the boundary condition from a known test solution, then use the collocation method with a piecewise cubic spline on equally spaced knots and collocation points to approximate the exact solution. It turns out that this alternative formulation is better conditioned than the version in Sect. 5.12. The results are validated through a comparison between the approximate and exact solutions.

5.13.2 Problem Statement

Domain boundary. The contour ∂S is the square parameterized by

$$
x1[t] = \begin{cases} t, & 0 \le t \le 1, \\ 1, & 1 < t \le 2, \\ 3-t, & 2 < t \le 3, \\ 0, & 3 < t \le 4, \end{cases} \qquad x2[t] = \begin{cases} 0, & 0 \le t \le 1, \\ -1+t, & 1 < t \le 2, \\ 1, & 2 < t \le 3, \\ 4-t & 3 < t \le 4. \end{cases} \tag{5.49}
$$

Its graph can be seen in Fig. 5.135.

Governing equations. The displacement u satisfies the boundary value problem

$$
Z_x \diamond u[x] = 0, \quad x \in S,
$$
$$
(\mathrm{T}u)[x] + \sigma u[x] = \mathscr{K}[x], \quad x \in \partial S.
$$

Test solution. The two components of the test solution u given by (5.3) are graphed in Fig. 5.130.

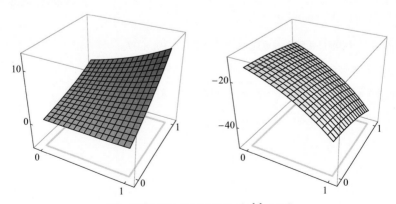

Fig. 5.130 The components of $u[x]$, $x \in S$.

Boundary data function. The function $\mathscr{K}$ is given by

$$(\mathrm{T}u)[x] + \sigma u[x] = \mathscr{K}[x], \quad x \in \partial S,$$

where, for computational purposes, we have chosen (without loss of generality) the particular 2×2 matrix

$$\sigma = \begin{pmatrix} 2 & 0 \\ 0 & 2 \end{pmatrix}.$$

The graphs of the two components of $\mathscr{K}$ are displayed in Figs. 5.131 and 5.132, in Cartesian coordinates and parametric form, respectively.

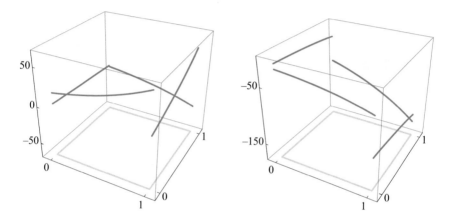

Fig. 5.131 The components of $\mathscr{K}[x]$ (Cartesian coordinates).

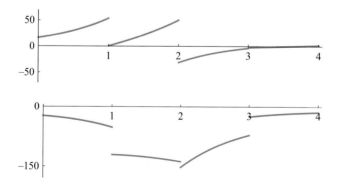

Fig. 5.132 The components of $\mathscr{K}[x[t]]$ (parametric form).

5.13.3 Solution Procedure

Method. We apply the alternative direct method, which reduces the problem to the boundary integral equation

$$\left(W_0 + \tfrac{1}{2}I\right)\varphi + V_0(\sigma\varphi) = V_0(\mathscr{K}) \quad \text{on } \partial S,$$

coded as

$$\oint_{\Gamma_{\mathrm{CPV}[x]}} P[x,y] \circ \varphi[y]\, d\Gamma_y + \oint_{\Gamma_{\mathrm{Weak}[x]}} D[x,y] \circ (\sigma\varphi[y])\, d\Gamma_y + \tfrac{1}{2}\,\varphi[x]$$

$$= \oint_{\Gamma_{\mathrm{Weak}[x]}} D[x,y] \circ \mathscr{K}[y]\, d\Gamma_y. \tag{5.50}$$

This equation is solved numerically to obtain an approximate density $\tilde{\varphi}$. In turn, $\tilde{\varphi}$ is used in the representation

$$u[x] = \oint_{\Gamma} D[x,y] \circ (\mathscr{K} - \sigma\varphi[y])\, d\Gamma_y - \oint_{\Gamma} P[x,y] \circ \varphi[y]\, d\Gamma_y, \quad x \in S, \tag{5.51}$$

to generate an approximate solution $\tilde{u}$ in S.

Density function. The exact density φ is the boundary stress vector Tu computed from the test solution u and given by (5.4); its components are graphed in Figs. 5.133 and 5.134.

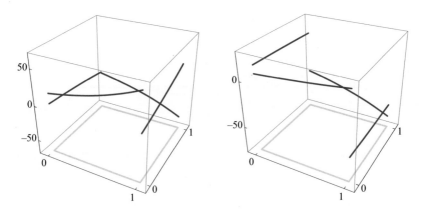

Fig. 5.133 The components of $\varphi[x]$ (Cartesian coordinates).

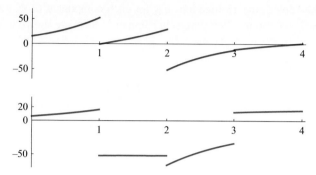

Fig. 5.134 The components of $\varphi[x[t]]$ (parametric form).

Numerical approximation. We compute $\tilde{\varphi}$ by the collocation method with a B-spline basis of elements $b_{i,j}$. Then the approximate density is sought in the form

$$\tilde{\varphi}[x[t]] = \begin{pmatrix} \sum_i \sum_j c_{1,i,j} b_{i,j}[t] \\ \sum_i \sum_j c_{2,i,j} b_{i,j}[t] \end{pmatrix}, \tag{5.52}$$

where the coefficients $c_{\alpha,i,j}$ are determined by substituting (5.52) in (5.50).

Since φ represents Tu on ∂S and the latter is discontinuous at the corners, the $b_{i,j}$ need to exhibit the same feature. In our case, we choose a piecewise cubic spline that is twice continuously differentiable at the secondary knots and discontinuous at the primary knot locations $t = 0, 1, 2, 3, 4$. In terms of the parameterization (5.49), the knots and the smoothness at their locations are specified as the set

$$\left\{0, 0, 0, \tfrac{1}{2}, 1, 1, 1, 1, \tfrac{3}{2}, 2, 2, 2, 2, \tfrac{5}{2}, 3, 3, 3, 3, \tfrac{7}{2}, 4, 4, 4, 4\right\}.$$

The knots are marked on the graph on the left in Fig. 5.135.

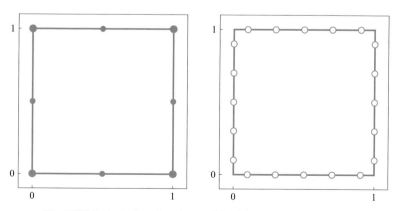

Fig. 5.135 Left: the knot locations on ∂S. Right: the collocation points.

The 8 knots determine 18 functions $b_{i,j}$ for each component of $\tilde{\varphi}$, generating a total of 36 basis functions. Their graphs in parametric form are shown in Fig. 5.136.

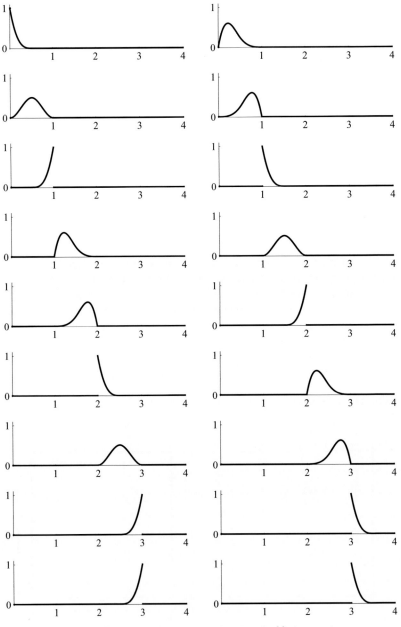

Fig. 5.136 The 20 B-spline basis functions $b_{i,j}[t]$, $0 \leq t \leq 4$.

To compute $\tilde{\varphi}$, we must choose an appropriate number of collocation points, which can be placed anywhere except at corners, because $\tilde{\varphi}$ is expected to be discontinuous there. Hence, we select collocation points that avoid the primary knots and are equally spaced. The set of 20 values of the parameter t at our selection is

$$\left\{ \tfrac{1}{10}, \tfrac{3}{10}, \tfrac{1}{2}, \tfrac{7}{10}, \tfrac{9}{10}, \tfrac{11}{10}, \tfrac{13}{10}, \tfrac{3}{2}, \tfrac{17}{10}, \tfrac{19}{10}, \tfrac{21}{10}, \tfrac{23}{10}, \tfrac{5}{2}, \tfrac{27}{10}, \tfrac{29}{10}, \tfrac{31}{10}, \tfrac{33}{10}, \tfrac{7}{2}, \tfrac{37}{10}, \tfrac{39}{10} \right\}.$$

These values are marked on the graph on the right in Fig. 5.135.

The same 20 collocation points are used for both components of $\tilde{\varphi}$, yielding a system of 40 constraining equations that enable us to compute the coefficients $c_{\alpha,i,j}$ in (5.52). The 40×36 matrix of this system has a condition number of 21.

5.13.4 Solution

Approximate density. The two components of $\tilde{\varphi}$ are graphed in Figs. 5.137 and 5.138. The latter also shows the locations of the primary and secondary knots.

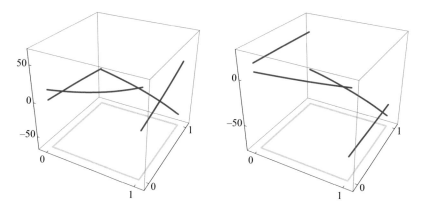

Fig. 5.137 The components of $\tilde{\varphi}[x]$ (Cartesian coordinates).

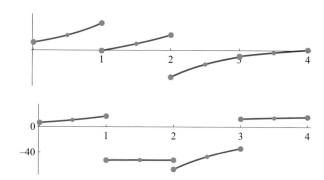

Fig. 5.138 The components of $\tilde{\varphi}[x[t]]$ (parametric form).

Approximate solution. We use $\tilde{\varphi}$ and $\mathscr{K}$ in (5.52) to compute an approximation $\tilde{u}$ to the exact solution u in S. The graphs of the two components of $\tilde{u}$ are shown in Fig. 5.139.

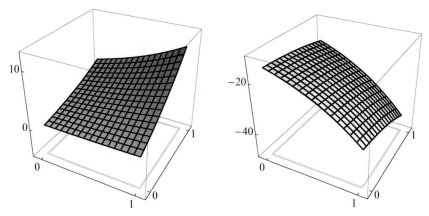

Fig. 5.139 The components of $\tilde{u}[x]$, $x \in S$.

Error analysis. To visualize the difference between the approximate and exact densities, in Fig. 5.140 we graphed $\tilde{\varphi} - \varphi$ relative to the maximum absolute value of φ on ∂S.

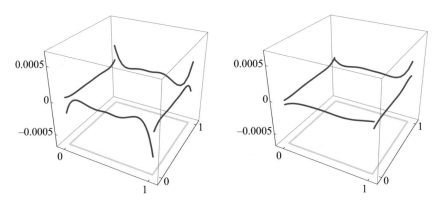

Fig. 5.140 The components of the relative error in $\tilde{\varphi}[x]$ (Cartesian coordinates).

The same was done in parametric form in Fig. 5.141, which also shows the primary and secondary knots and indicates that the relative error in the two components of $\tilde{\varphi}$ is about 0.02%. With such a small error on the boundary, we expect an equally good agreement in the interior of the domain.

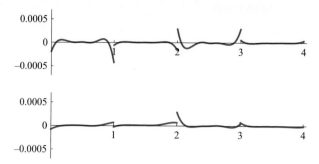

Fig. 5.141 The components of the relative error in $\tilde{\varphi}[x[t]]$ (parametric form).

5.14 Remarks. (i) The amount of computation needed when we use (5.50) or (5.46) is about the same. This is expected because of the similarity in the structure of these equations. Also, the overall relative accuracy in both cases is similar (considering that the former calculates u and the latter calculates Tu on the boundary). The accuracy for (5.50) is slightly better, which is to be expected because the B-spline basis for it has more elements. However, what is not expected is that the condition number of the coefficient matrix resulting from the use of the (5.50) is 21, significantly smaller than the condition number of 245 when we use equation (5.46). Further work is required to see if this trend is confirmed and, thus, we are justified to use of the alternative direct method in preference to the 'classical' direct method.

(ii) The set of coefficients $\{c_{1,i,j}, c_{2,i,j}\}$ in (5.52), computed with the B-spline basis shown in Fig. 5.136, is

$$\{\{15.1, 18.6, 27.2, 42., 52.2, -0.932, 3.12, 11.6, 22.7, 29.4, -52.,$$
$$-41.1, -25.3, -16.2, -12.5, -11.2, -8.45, -4.02, -0.707, 0.865\},$$
$$\{6.11, 7.33, 10., 13.9, 16.3, -52.2, -52., -52., -52.1, -52., -66.4,$$
$$-58.1, -45.2, -37., -33.7, 12.5, 13.2, 14.1, 14.7, 15.1\}\}.$$

5.14 Dirichlet Problem in a Square: Classical Indirect Method

5.14.1 Summary

This example illustrates the application of the classical indirect method to the Dirichlet boundary value problem in a square domain. We compute the boundary condition from a known test solution, then use the collocation method with a piecewise cubic spline on equally spaced knots and collocation points to approximate the exact solution. The results for the unknown density are validated by means of two different techniques.

5.14.2 Problem Statement

Domain boundary. ∂S is the square parameterized by

$$
x1[t] = \begin{cases} t, & 0 \le t \le 1, \\ 1, & 1 < t \le 2, \\ 3-t, & 2 < t \le 3, \\ 0, & 3 < t \le 4, \end{cases} \qquad x2[t] = \begin{cases} 0, & 0 \le t \le 1, \\ t-1, & 1 < t \le 2, \\ 1, & 2 < t \le 3, \\ 4-t & 3 < t \le 4. \end{cases} \tag{5.53}
$$

Its graph can be seen in Fig. 5.145.

Governing equations. The displacement u is the solution of the boundary value problem

$$
\begin{aligned}
Z_x \diamond u[x] &= 0, & x \in S, \\
u[x] &= \mathscr{P}[x], & x \in \partial S.
\end{aligned}
$$

Test solution. The two components of the test solution u given by (5.3) are graphed in Fig. 5.142.

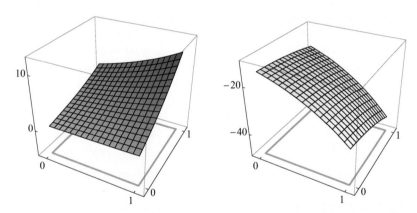

Fig. 5.142 The components of $u[x]$, $x \in S$.

Boundary data function. The function $\mathscr{P}$ is computed from the test solution u as

$$
\mathscr{P}[x] = u[x], \qquad x \in \partial S.
$$

The graphs of its two components are displayed in Figs. 5.143 and 5.144, in Cartesian coordinates and parametric form, respectively.

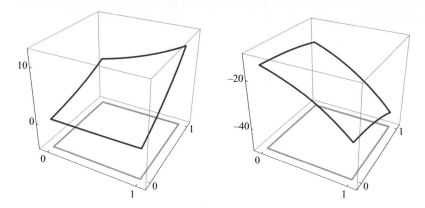

Fig. 5.143 The components of $\mathscr{P}[x]$ (Cartesian coordinates).

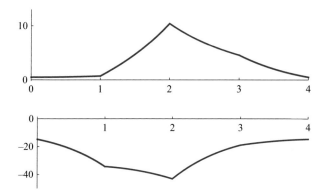

Fig. 5.144 The components of $\mathscr{P}[x[t]]$ (parametric form).

5.14.3 Solution Procedure

Method. We apply the classical indirect method, which reduces the problem to the boundary integral equation

$$\left(W_0 - \tfrac{1}{2}I\right)\varphi = \mathscr{P} \quad \text{on } \partial S,$$

coded as

$$\oint_{\Gamma_{\mathrm{CPV}[x]}} P[x,y] \circ \varphi[y]\, d\Gamma_y - \tfrac{1}{2}\,\varphi[x] = \mathscr{P}[x]. \tag{5.54}$$

This equation is solved numerically to obtain an approximate density $\tilde{\varphi}$. In turn, $\tilde{\varphi}$ is used in the representation

$$u[x] = \oint_{\Gamma} P[x,y] \circ \varphi[y] \, d\Gamma_y, \quad x \in S \tag{5.55}$$

to generate an approximate solution $\tilde{u}$ in S.

Density function. The exact density φ does not represent either u or Tu on ∂S, which makes validation of the results somewhat more difficult.

Numerical approximation. We compute $\tilde{\varphi}$ by the collocation method with a B-spline basis of elements $b_{i,j}$. Then the approximate density is sought in the form

$$\tilde{\varphi}[x[t]] = \begin{pmatrix} \sum_i \sum_j c_{1,i,j} b_{i,j}[t] \\ \sum_i \sum_j c_{2,i,j} b_{i,j}[t] \end{pmatrix}, \tag{5.56}$$

where the coefficients $c_{\alpha,i,j}$ are determined by substituting (5.56) in (5.54).

The construction of a spline basis requires that the smoothness of the spline at the knots must be the same as, or weaker than, that of the function being approximated. Therefore, since we do not know the behavior of φ at the corners, we have to work with a spline that is discontinuous at those four points. In our case, we choose a piecewise cubic spline that is twice continuously differentiable at the secondary knots and discontinuous at the primary knot locations $t = 0, 1, 2, 3, 4$. In terms of the parameterization (5.53), the knots and the smoothness at their locations are specified as the set

$$\left\{ 0, 0, 0, 0, \tfrac{1}{3}, \tfrac{2}{3}, 1, 1, 1, 1, \tfrac{4}{3}, \tfrac{5}{3}, 2, 2, 2, 2, \tfrac{7}{3}, \tfrac{8}{3}, 3, 3, 3, 3, \tfrac{10}{3}, \tfrac{11}{3}, 4, 4, 4, 4 \right\}.$$

The knots are marked on the graph on the left in Fig. 5.145.

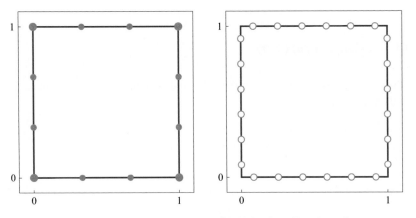

Fig. 5.145 Left: the knot locations on ∂S. Right: the collocation points.

The 13 knots determine 24 functions $b_{i,j}$ for each component of $\tilde{\varphi}$, generating 48 basis functions. The graphs of the $b_{i,j}$ in parametric form are shown in Fig. 5.146.

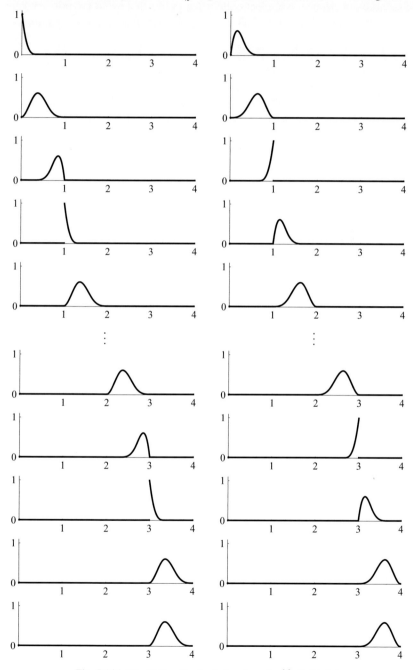

Fig. 5.146 The 48 B-spline basis functions $b_{i,j}[t]$, $0 \le t \le 4$.

To compute $\tilde{\varphi}$, we must choose an appropriate number of collocation points, which can be placed anywhere except at the corners $t = 0, 1, 2, 3, 4$, where φ might be discontinuous. Hence, we select collocation points that avoid the primary knots and are equally spaced. The set of 24 values of the parameter t at our selection is

$$\left\{ \frac{1}{12}, \frac{1}{4}, \frac{5}{12}, \frac{7}{12}, \frac{3}{4}, \frac{11}{12}, \frac{13}{12}, \frac{5}{4}, \frac{17}{12}, \frac{19}{12}, \frac{7}{4}, \frac{23}{12}, \right.$$
$$\left. \frac{25}{12}, \frac{9}{4}, \frac{29}{12}, \frac{31}{12}, \frac{11}{4}, \frac{35}{12}, \frac{37}{12}, \frac{13}{4}, \frac{41}{12}, \frac{43}{12}, \frac{15}{4}, \frac{47}{12} \right\}.$$

These values are marked on the graph on the right in Fig. 5.145.

The same 24 points are used for both components of $\tilde{\varphi}$, yielding a system of 48 constraining equations that enable us to compute the numerical coefficients $c_{\alpha,i,j}$ in (5.56). The 48×48 matrix of this system has a condition number of 41.

5.14.4 Solution

Approximate density. The two components of $\tilde{\varphi}$ are graphed in Figs. 5.147 and 5.148. These graphs strongly suggests that the density φ is continuous at the primary knot locations $t = 0, 1, 2, 3, 4$.

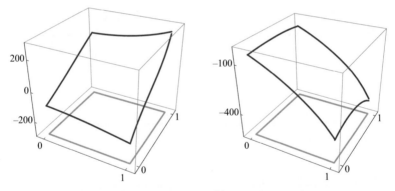

Fig. 5.147 The components of $\tilde{\varphi}[x]$ (Cartesian coordinates).

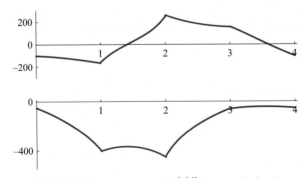

Fig. 5.148 The components of $\tilde{\varphi}[x[t]]$ (parametric form).

Approximate solution. We use $\tilde{\varphi}$ and $\mathscr{P}$ in (5.55) to compute an approximation $\tilde{u}$ to the exact solution u in S. The graphs of the two components of $\tilde{u}$, shown in Fig. 5.149, are virtually indistinguishable from those of u in Fig. 5.142.

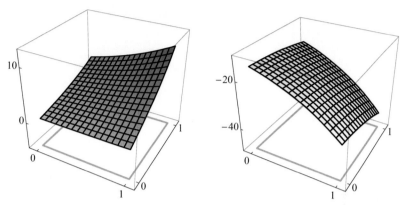

Fig. 5.149 The components of $\tilde{u}[x]$, $x \in S$.

Error analysis. A direct comparison between $\tilde{\varphi}$ and φ is not possible because the latter is unknown. However, we can do this indirectly by using the fact that φ satisfies (see Sect. 5.14.3)

$$\left(W_0 - \tfrac{1}{2}I\right)\varphi = \mathscr{P} \quad \text{on } \partial S.$$

The approximate density $\tilde{\varphi}$ replaced in that equation produces an approximation $\tilde{\mathscr{P}}$ to $\mathscr{P}$. The difference $\tilde{\mathscr{P}} - \mathscr{P}$ relative to the maximum absolute value of $\mathscr{P}$ on ∂S is graphed in Figs. 5.150 and 5.151. The latter shows that the relative error in the two components of $\tilde{\mathscr{P}}$ is less than 1%, which confirms the validity of our computation.

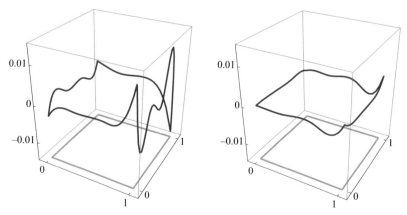

Fig. 5.150 The components of the relative error in $\tilde{\mathscr{P}}[x]$ (Cartesian coordinates).

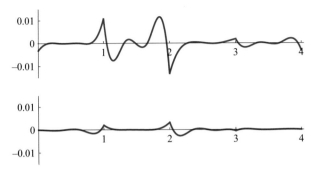

Fig. 5.151 The components of the relative error in $\tilde{\mathscr{P}}[x[t]]$ (parametric form).

A second way of comparing $\tilde{\varphi}$ and φ is by looking at the difference between $\tilde{u}$ in S, generated by (5.56) with φ replaced by $\tilde{\varphi}$, and u in S, generated by the same formula with the exact density φ. The difference $\tilde{u} - u$ relative to the maximum absolute value of u in S, graphed in Fig. 5.152, shows good agreement between the computed and exact solutions, so, once again, we are confident that our numerical results are valid.

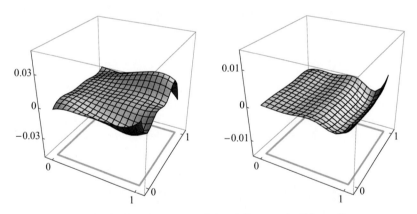

Fig. 5.152 The components of the relative error in $\tilde{u}[x]$, $x \in S$.

5.15 Remarks. (i) As mentioned earlier, the graphs of $\tilde{\varphi}$ strongly suggest that the exact density φ is continuous at the corners of the domain, but that its derivatives are not. This feature should be taken into account when we construct a spline basis.

(ii) Computationally, the classical indirect method appears to be more sensitive to small errors than the direct method for similar domains with similar boundary conditions. As a result, the use of the former requires greater attention to be paid to computational accuracy.

(iii) The set of coefficients $\{c_{1,i,j}, c_{2,i,j}\}$ in (5.56), computed with the B-spline basis shown in Fig. 5.146, is

$$
\begin{aligned}
\{\{&-100., -102., -111., -133., -153., -164., -158., -96.8, \\
&-20.4, 71., 162., 247., 259., 234., 198., 165., \\
&156., 154., 156., 131., 70., -22., -78.2, -101.\}, \\
\{&-52.4, -74.3, -127., -229., -326., -397., -404., -383., \\
&-133., -360., -374., -419., -455., -442., -347., -237., \\
&-79.6, -58.4, -60.8, -52.2, -43.2, -42.9, -48.8, -54.1\}\}.
\end{aligned}
$$

5.15 Dirichlet Problem in an Asymmetric Domain: Classical Indirect Method

5.15.1 Summary

This example illustrates the application of the classical indirect method to the Dirichlet boundary value problem in a domain with a smooth asymmetric boundary. We compute the boundary condition from a known test solution, then use the collocation method with a piecewise linear spline on equally spaced knots and collocation points to approximate the exact solution. The results for the unknown density are validated indirectly by reference to the boundary data function.

5.15.2 Problem Statement

Domain boundary. The contour ∂S is the curve parameterized by

$$
\begin{aligned}
x1[t] &= \tfrac{1}{10}\left(7 + 8\mathrm{Cos}[\pi t] + 4\mathrm{Cos}[2\pi t]\right), \\
x2[t] &= \tfrac{1}{50}\left(25 + 2\mathrm{Cos}[3\pi t] + 20\mathrm{Sin}[\pi t]\right),
\end{aligned}
\qquad 0 \le t \le 2.
\qquad (5.57)
$$

Its graph can be seen in Fig. 5.156.

Governing equations. The displacement u is the solution of the boundary value problem

$$
\begin{aligned}
Z_x \diamond u[x] &= 0, \quad x \in S, \\
u[x] &= \mathscr{P}[x], \quad x \in \partial S.
\end{aligned}
$$

Test solution. The two components of the test solution u given by (5.3) are graphed in Fig. 5.153.

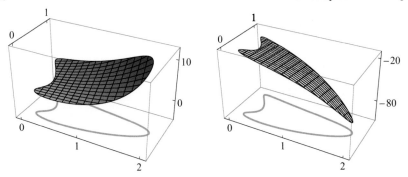

Fig. 5.153 The components of $u[x]$, $x \in S$.

Boundary data function. The function $\mathscr{P}$ is computed from the test solution u as

$$\mathscr{P}[x] = u[x], \quad x \in \partial S.$$

The graphs of its two components are displayed in Figs. 5.154 and 5.155, in Cartesian coordinates and parametric form, respectively.

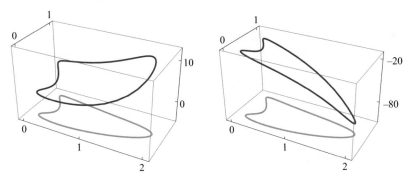

Fig. 5.154 The components of $\mathscr{P}[x]$ (Cartesian coordinates).

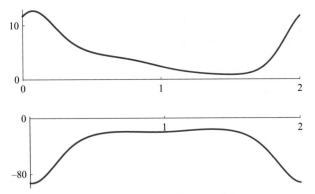

Fig. 5.155 The components of $\mathscr{P}[x[t]]$ (parametric form).

5.15.3 Solution Procedure

Method. We apply the classical indirect method, which reduces the problem to the boundary integral equation

$$\left(W_0 - \tfrac{1}{2}I\right)\varphi = \mathscr{P} \quad \text{on } \partial S,$$

coded as

$$\oint_{\Gamma_{\text{CPV}[x]}} P[x,y] \circ \varphi[y]\, d\Gamma_y - \tfrac{1}{2}\varphi[x] = \mathscr{P}[x]. \tag{5.58}$$

This equation is solved numerically to obtain an approximate density $\tilde{\varphi}$. In turn, $\tilde{\varphi}$ is used in the representation

$$u[x] = \oint_{\Gamma} P[x,y] \circ \varphi[y]\, d\Gamma_y, \quad x \in S \tag{5.59}$$

to generate an approximate solution $\tilde{u}$ in S.

Density function. The exact density φ does not represent either u or Tu on ∂S, which makes validation of the results somewhat more difficult.

Numerical approximation. We compute $\tilde{\varphi}$ by the collocation method with a B-spline basis of elements $b_{i,j}$. Then the approximate density is sought in the form

$$\tilde{\varphi}[x[t]] = \begin{pmatrix} \sum_i \sum_j c_{1,i,j} b_{i,j}[t] \\ \sum_i \sum_j c_{2,i,j} b_{i,j}[t] \end{pmatrix}, \tag{5.60}$$

where the coefficients $c_{\alpha,i,j}$ are determined by substituting (5.60) in (5.58).

The construction of a spline basis requires the smoothness of the spline at the knots to be the same as, or weaker than, that of the approximated function. Since the parameterization of ∂S is smooth, we expect the exact density φ to be smooth on ∂S as well, so we construct a spline with the same feature. In our case, we choose a piecewise linear spline continuous at all the knots. In terms of the parameterization (5.57), the knots and the smoothness at their locations are specified as the set

$$\left\{ 0, \tfrac{1}{14}, \tfrac{1}{7}, \tfrac{3}{14}, \tfrac{2}{7}, \tfrac{5}{14}, \tfrac{3}{7}, \tfrac{1}{2}, \tfrac{4}{7}, \tfrac{9}{14}, \tfrac{5}{7}, \tfrac{11}{14}, \tfrac{6}{7}, \tfrac{13}{14}, 1, \right.$$
$$\left. \tfrac{15}{14}, \tfrac{8}{7}, \tfrac{17}{14}, \tfrac{9}{7}, \tfrac{19}{14}, \tfrac{10}{7}, \tfrac{3}{2}, \tfrac{11}{7}, \tfrac{23}{14}, \tfrac{12}{7}, \tfrac{25}{14}, \tfrac{13}{7}, \tfrac{27}{14}, 2 \right\}.$$

The knots are marked on the graph on the left in Fig. 5.156.

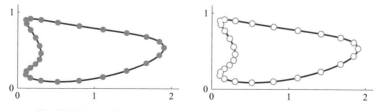

Fig. 5.156 Left: the knot locations on ∂S. Right: the collocation points.

The 28 knots determine 28 functions $b_{i,j}$ for each component of $\tilde{\varphi}$, generating 56 basis functions whose graphs, in parametric form, are shown in Fig. 5.157.

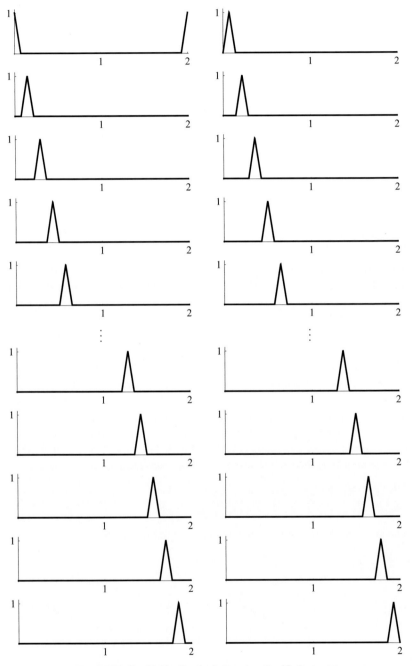

Fig. 5.157 The 28 B-spline basis functions $b_{i,j}[t]$, $0 \leq t \leq 2$.

To compute $\tilde{\varphi}$, we must choose an appropriate number of collocation points. The expected smoothness of φ allows us to place these points almost anywhere. The set of 28 equally-spaced values of the parameter t at our selection is

$$\left\{0, \frac{1}{14}, \frac{1}{7}, \frac{3}{14}, \frac{2}{7}, \frac{5}{14}, \frac{3}{7}, \frac{1}{2}, \frac{4}{7}, \frac{9}{14}, \frac{5}{7}, \frac{11}{14}, \frac{6}{7}, \frac{13}{14}, 1,\right.$$
$$\left.\frac{15}{14}, \frac{8}{7}, \frac{17}{14}, \frac{9}{7}, \frac{19}{14}, \frac{10}{7}, \frac{3}{2}, \frac{11}{7}, \frac{23}{14}, \frac{12}{7}, \frac{25}{14}, \frac{13}{7}, \frac{27}{14}\right\}.$$

These values are marked on the graph on the right in Fig. 5.156.

The same 28 points are used for both components of $\tilde{\varphi}$, yielding a system of 56 constraining equations that enable us to compute the coefficients $c_{\alpha,i,j}$ in (5.60). The 56×56 matrix of this system has a condition number of 27.

5.15.4 Solution

Approximate density. The components of $\tilde{\varphi}$ are graphed in Figs. 5.158 and 5.159.

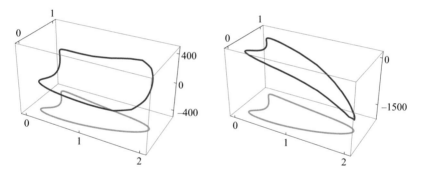

Fig. 5.158 The components of $\tilde{\varphi}[x]$ (Cartesian coordinates).

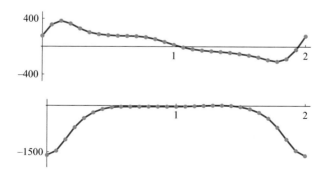

Fig. 5.159 The components of $\tilde{\varphi}[x[t]]$ (parametric form).

Error analysis. A direct comparison between $\tilde{\varphi}$ and φ is not possible because the latter is unknown. However, we can do this indirectly by using the fact that φ is the exact solution of the equation

$$\left(W_0 - \tfrac{1}{2} I\right) \varphi = \mathscr{P} \quad \text{on } \partial S.$$

The approximate density $\tilde{\varphi}$ replaced in that equation produces an approximation $\tilde{\mathscr{P}}$ to $\mathscr{P}$. Both functions $\tilde{\mathscr{P}}$ and $\mathscr{P}$ are graphed in Figs. 5.160 and 5.161.

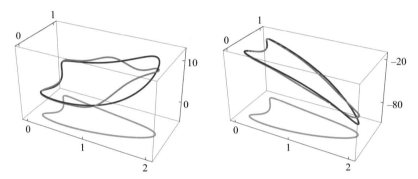

Fig. 5.160 The components of $\tilde{\mathscr{P}}[x]$ (orange line) and $\mathscr{P}[x]$ (blue line) in Cartesian coordinates.

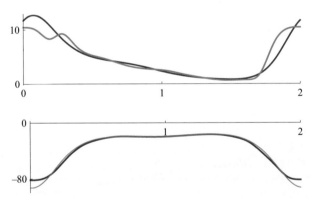

Fig. 5.161 The components of $\tilde{\mathscr{P}}[x[t]]$ (orange line) and $\mathscr{P}[x[t]]$ (blue line) in parametric form.

We note that the agreement between the two is not very good. The first component of $\tilde{\mathscr{P}}$ in Fig. 5.160 is twisted on the right-hand side of the domain, which is also where the parameterization begins and ends (see Fig. 5.161). Such behavior is characteristic of the errors produced by the classical indirect method and can become quite severe. We expected this to happen because we chose equal spacing of the parameter values for both the spline knots and collocation points, and not for their geometric positions to the boundary. As Fig. 5.156 shows, the density of these points is high on the left segment of ∂S and low on the right one. Subsequent calculations, not shown here, show that a more uniform redistribution of the spline knots and collocation points along the boundary rather than with respect to their parameter values yields significantly better results.

5.16 Remarks. (i) In the classical indirect method, the collocation points must be chosen carefully. Numerical experimentation shows that placing them midway between the spline knots when using piecewise linear splines produces an ill-conditioned problem. To avoid this, in our example we chose the collocation points at the knot locations.

(ii) In Fig. 5.157, the first basis function was modified to exhibit periodicity and thus be consistent with the expected continuity of φ.

(iii) The set of coefficients $\{c_{1,i,j}, c_{2,i,j}\}$ in (5.60), computed with the B-spline basis shown in Fig. 5.157, is

$$\{\{167., 359., 416., 358., 267., 203., 175., 160., 150., 148., 147.,$$
$$135., 108., 68., 23.6, -13.9, -40., -56.6, -67.7, -76.6, -88.,$$
$$-105., -126., -157., -203., -242., -214., -65.3\}, \{-1760.,$$
$$-1580., -1150., -709., -391., -204., -94.8, -37.8, -24.3,$$
$$-25.4, -22.5, -17.5, -15.4, -15.4, -14.1, -8.68, 1.82, 14., 22.6,$$
$$24., 14., -19.4, -89.1, -201., -387., -699., -1130., -1570.\}\}.$$

5.16 Neumann Problem in an Ellipse: Classical Indirect Method

5.16.1 Summary

This example illustrates the application of the classical indirect method to a Neumann problem in a domain with a smooth boundary. We compute the boundary condition from a known test solution, then use the collocation method with a piecewise cubic spline on equally spaced knots and collocation points to approximate the exact density. We also discuss the occurrence of ill-conditioning and indicate a technique for mitigating its effect. Since the exact density is unknown, the results are validated through an indirect procedure.

5.16.2 Problem Statement

Domain boundary. ∂S is the ellipse parameterized by

$$x1[t] = 1 + \text{Cos}[\pi t], \quad x2[t] = \tfrac{1}{2}\text{Sin}[\pi t] + \tfrac{1}{2}, \quad 0 \le t \le 2. \tag{5.61}$$

Its graph can be seen in Fig. 5.165.

Governing equations. The displacement u satisfies the boundary value problem

$$Z_x \diamond u[x] = 0, \quad x \in S,$$
$$(\text{Tu})[x] = T_x \diamond u[x] = \mathcal{Q}[x], \quad x \in \partial S.$$

Test solution. The two components of the test solution u given by (5.3) are graphed in Fig. 5.162.

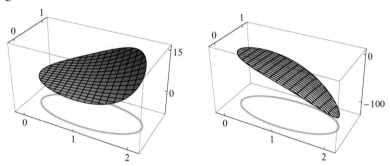

Fig. 5.162 The components of $u[x]$, $x \in S$.

Boundary data function. The function $\mathcal{Q} = \text{Tu}$ is given by (5.4). The graphs of its two components are displayed in Figs. 5.163 and 5.164, in Cartesian coordinates and parametric form, respectively.

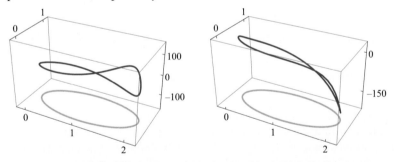

Fig. 5.163 The components of $\mathcal{Q}[x]$ (Cartesian coordinates).

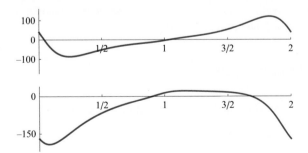

Fig. 5.164 The components of $\mathcal{Q}[x[t]]$ (parametric form).

5.16.3 Solution Procedure

Method. We apply the classical indirect method, which reduces the problem to the boundary integral equation

$$\left(W_0^* + \tfrac{1}{2}I\right)\psi = \mathcal{Q} \quad \text{on } \partial S,$$

coded as

$$\oint_{\Gamma_{\mathrm{CPV}[x]}} (T_x \diamond D[x,y]) \circ \psi[y]\, d\Gamma_y + \tfrac{1}{2}\,\psi[x] = \mathcal{Q}[x]. \tag{5.62}$$

This equation is solved numerically to obtain an approximate density $\tilde{\psi}$. In turn, $\tilde{\psi}$ is used in the representation

$$u[x] = \oint_{\Gamma} D[x,y] \circ \psi[y]\, d\Gamma_y, \quad x \in S \tag{5.63}$$

to generate an approximate solution $\tilde{u}$ in S.

Density function. The exact density ψ does not represent either u or $\mathrm{T}u$ on ∂S, which makes validation of the results somewhat more difficult.

Numerical approximation. We compute $\tilde{\psi}$ by the collocation method with a B-spline basis of elements $b_{i,j}$. Then the approximate density is sought in the form

$$\tilde{\psi}[x[t]] = \begin{pmatrix} \sum_i \sum_j c_{1,i,j} b_{i,j}[t] \\ \sum_i \sum_j c_{2,i,j} b_{i,j}[t] \end{pmatrix}, \tag{5.64}$$

where the coefficients $c_{\alpha,i,j}$ are determined by substituting (5.64) in (5.62).

The construction of a spline basis requires the smoothness of the spline at the knots to be the same as, or weaker than, that of the approximated function. Since the parameterization (5.61) is smooth, we expect ψ to be equally smooth on ∂S, so we may construct an approximate density $\tilde{\psi}$ with any degree of smoothness. Therefore, to ensure maximum smoothness, we choose a piecewise cubic spline twice continuously differentiable at the knots and modified to enforce periodicity and continuity at the knots corresponding to $t = 0, 2$. In terms of the parameterization (5.61), the knots and the smoothness at their locations are specified as the set

$$\left\{0, \tfrac{1}{6}, \tfrac{1}{3}, \tfrac{1}{2}, \tfrac{2}{3}, \tfrac{5}{6}, 1, \tfrac{7}{6}, \tfrac{4}{3}, \tfrac{3}{2}, \tfrac{5}{3}, \tfrac{11}{6}, 2\right\}.$$

The knots are marked on the graph on the left in Fig. 5.165.

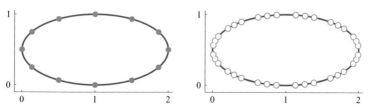

Fig. 5.165 Left: the knot locations on ∂S. Right: the collocation points.

The 12 knots determine 12 functions $b_{i,j}$ for each component of $\tilde{\psi}$, generating a total of 24 basis functions. The graphs of the $b_{i,j}$ in parametric form are shown in Fig. 5.166.

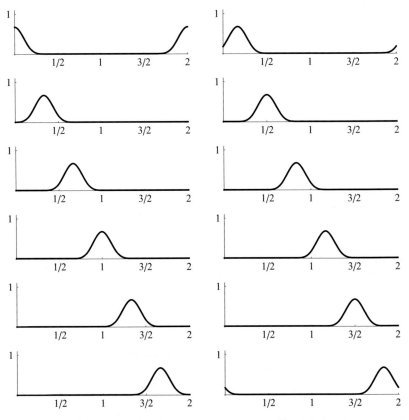

Fig. 5.166 The 12 B-spline basis functions $b_{i,j}[t]$, $0 \le t \le 2$.

To compute $\tilde{\psi}$, we must choose an appropriate number of collocation points. Owing to the smoothness of ∂S, these points can be placed almost anywhere. However, computational experiments (not shown here) indicate that the typical choice for these points generates a singular coefficient matrix. Consequently, to help with the computational stability of the method, we have augmented the original set with additional locations, which gives rise to an overdetermined system. We select collocation points that avoid the knots and are equally spaced. The set of 36 values of the parameter t at our selection is

$$\left\{ \frac{1}{24}, \frac{1}{12}, \frac{1}{8}, \frac{5}{24}, \frac{1}{4}, \frac{7}{24}, \frac{3}{8}, \frac{5}{12}, \frac{11}{24}, \frac{13}{24}, \frac{7}{12}, \frac{5}{8}, \frac{17}{24}, \frac{3}{4}, \frac{19}{24}, \frac{7}{8}, \frac{11}{12}, \frac{23}{24}, \right.$$
$$\left. \frac{25}{24}, \frac{13}{12}, \frac{9}{8}, \frac{29}{24}, \frac{5}{4}, \frac{31}{24}, \frac{11}{8}, \frac{17}{12}, \frac{35}{24}, \frac{37}{24}, \frac{19}{12}, \frac{13}{8}, \frac{41}{24}, \frac{7}{4}, \frac{43}{24}, \frac{15}{8}, \frac{23}{12}, \frac{47}{24} \right\}.$$

These values are marked on the graph on the right in Fig. 5.165.

The same 36 points are used for both components of the approximation $\tilde{\psi}$, yielding a system of 72 constraining equations that enable us to compute the coefficients $c_{\alpha,i,j}$ in (5.64). The 72×24 matrix of this overdetermined system has a condition number of 12.

5.16.4 Solution

Approximate density. The two components of $\tilde{\psi}$ are graphed in Figs. 5.167 and 5.168. These graphs strongly suggests that our conjecture regarding the continuity of the exact density ψ on ∂S is correct.

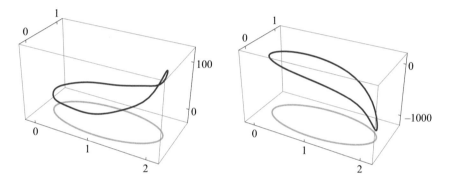

Fig. 5.167 The components of $\tilde{\psi}[x]$ (Cartesian coordinates).

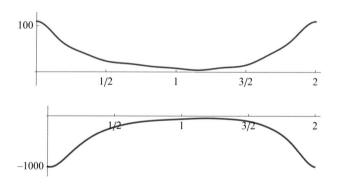

Fig. 5.168 The components of $\tilde{\psi}[x[t]]$ (parametric form).

Approximate solution. We use $\tilde{\psi}$ in (5.63) to compute an approximation $\tilde{u}$ to the exact solution u in S. The graphs of the components of $\tilde{u}$ are shown in Fig. 5.169. There is virtually no visual distinction between $\tilde{u}$ in Fig. 5.169 and u in Fig. 5.162.

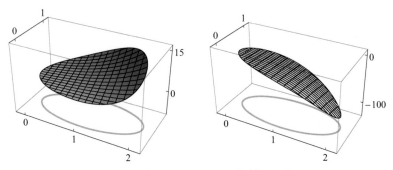

Fig. 5.169 The components of $\tilde{u}[x]$, $x \in S$.

Error analysis. A direct comparison between $\tilde{\psi}$ and ψ is not possible because the latter is unknown. However, we can do this indirectly by using the fact that u is computed from ψ by means of (5.63). The approximate density $\tilde{\psi}$ replaced in the restriction of this formula to the boundary produces an approximation $\tilde{u}$ to u on ∂S. The difference $\tilde{u}[x] - u[x]$, $x \in \partial S$, relative to the maximum absolute value of u on ∂S is graphed in Figs. 5.170 and 5.171. The latter shows that the relative error in the two components of $\tilde{u}$ on the boundary is less than 1%, which confirms the validity of our computation.

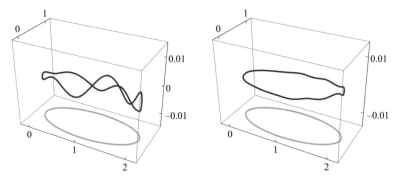

Fig. 5.170 The components of the relative error in $\tilde{u}[x]$, $x \in \partial S$ (Cartesian coordinates).

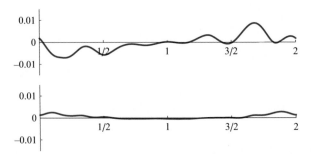

Fig. 5.171 The components of the relative error in $\tilde{u}[x[t]]$, $0 \le t \le 2$ (parametric form).

5.17 Remarks. (i) Computationally, the classical indirect method appears to be more sensitive to small errors than the direct method for similar domains with similar boundary conditions. As a result, the use of the former requires greater attention to be paid to computational accuracy. Here, we used surplus collocation points to diminish the instability inherent in this technique.

(ii) The set of coefficients $\{c_{1,i,j}, c_{2,i,j}\}$ in (5.64), computed with the B-spline basis shown in Fig. 5.166, is

$$\{\{133., 60., 42.7, 20., 19.2, 10.6, 9.37, 1.19, 11.7, 10.,$$
$$39.2, 61.6\}, \{-1200., -734., -377., -207., -116.,$$
$$-81.2, -63., -37.7, -53.4, -88.5, -220., -494.\}\}.$$

5.17 Neumann Problem in a Square: Classical Indirect Method

5.17.1 Summary

This example illustrates the application of the classical indirect method to a Neumann problem in a square domain, where the boundary condition is computed from a known test solution. We use the collocation method with a piecewise cubic spline on unequally spaced knots and collocation points to construct an approximation of the original solution from the computed boundary data function. The validation process, carried out through a comparison between the exact and approximate solutions, indicates that the classical indirect method is not recommended for use in domains with corners.

5.17.2 Problem Statement

Domain boundary. ∂S is the square parameterized by

$$x1[t] = \begin{cases} t, & 0 \le t \le 1, \\ 1, & 1 < t \le 2, \\ 3-t, & 2 < t \le 3, \\ 0, & 3 < t \le 4, \end{cases} \quad x2[t] = \begin{cases} 0, & 0 \le t \le 1, \\ t-1, & 1 < t \le 2, \\ 1, & 2 < t \le 3, \\ 4-t & 3 < t \le 4. \end{cases} \tag{5.65}$$

Its graph can be seen in Fig. 5.175.

Governing equations. The displacement u is the solution of the boundary value problem

$$Z_x \diamond u[x] = 0, \quad x \in S,$$
$$(Tu)[x] = T_x \diamond u[x] = \mathscr{Q}[x], \quad x \in \partial S.$$

Test solution. The two components of u given by (5.3) are graphed in Fig. 5.172.

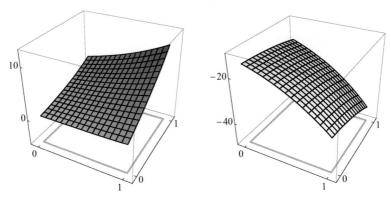

Fig. 5.172 The components of $u[x]$, $x \in S$.

Boundary data function. The function $\mathcal{Q} = \mathrm{T}u$ is given by (5.4). The graphs of its two components are displayed in Figs. 5.173 and 5.174, in Cartesian coordinates and parametric form, respectively.

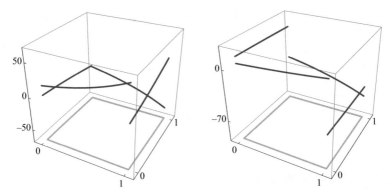

Fig. 5.173 The components of $\mathcal{Q}[x]$ (Cartesian coordinates).

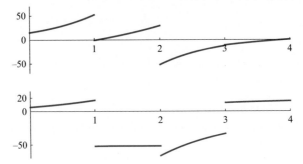

Fig. 5.174 The components of $\mathcal{Q}[x[t]]$ (parametric form).

5.17.3 Solution Procedure

Method. We apply the classical indirect method, which reduces the problem to the boundary integral equation

$$\left(W_0^* + \tfrac{1}{2}I\right)\psi = \mathcal{Q} \quad \text{on } \partial S,$$

coded as

$$\oint_{\Gamma_{\text{CPV}[x]}} (T_x \diamond D[x,y]) \circ \psi[y] \, d\Gamma_y + \tfrac{1}{2}\,\psi[x] = \mathcal{Q}[x]. \tag{5.66}$$

This equation is solved numerically to obtain an approximate density $\tilde{\psi}$. In turn, $\tilde{\psi}$ is used in the representation

$$u[x] = \oint_{\Gamma} D[x,y] \circ \psi[y] \, d\Gamma_y, \quad x \in S \tag{5.67}$$

to generate an approximate solution $\tilde{u}$ in S.

Density function. The exact density ψ does not represent either u or Tu on ∂S, which makes validation of the results somewhat more difficult.

Numerical approximation. We compute $\tilde{\psi}$ by the collocation method with a B-spline basis of elements $b_{i,j}$. Consequently, the approximate density is sought in the form

$$\tilde{\psi}[x[t]] = \begin{pmatrix} \sum_i \sum_j c_{1,i,j} b_{i,j}[t] \\ \sum_i \sum_j c_{2,i,j} b_{i,j}[t] \end{pmatrix}, \tag{5.68}$$

where the numerical coefficients $c_{\alpha,i,j}$ are determined by substituting (5.68) in (5.66).

The construction of a spline basis requires that the smoothness of the spline at the knots must be the same as, or weaker than, that of the function being approximated. Therefore, since we do not know the behavior of ψ at the corners, we have to work with a spline that is discontinuous at those four points. In this case, we choose a piecewise cubic spline that is twice continuously differentiable at the secondary knots and discontinuous at the primary knot locations $t = 0, 1, 2, 3, 4$.

According to the calculations below, ψ appears to have a singularity at the corners, so the collocation method would perform very poorly if we were to use a typical uniform distribution of spline knots. Instead, we choose knots and collocation points that are densely clustered about the parametric values $t = 0, 1, 2, 3, 4$ corresponding to the four corners of the boundary. This type of choice helps us judge the behavior of ψ at those points. In terms of the parameterization (5.65), the knots and the smoothness at their locations are specified as the set

$\{0, 0, 0, 0, 0.00302, 0.0141, 0.107, 0.5, 0.893, 0.986, 0.997, 1, 1, 1, 1, 1.003,$
$1.014, 1.107, 1.5, 1.893, 1.986, 1.997, 2, 2, 2, 2, 2.003, 2.014, 2.107, 2.5,$
$2.893, 2.986, 2.997, 3, 3, 3, 3, 3.003, 3.014, 3.107, 3.5, 3.893, 3.986, 3.997,$
$4, 4, 4, 4\}.$

The knots are marked on the graph on the left in Fig. 5.175. We draw attention to the fact that, owing to the closeness of the knots near the corners, 16 of the 28 secondary knot locations are hidden under the large points designating the four primary knot locations.

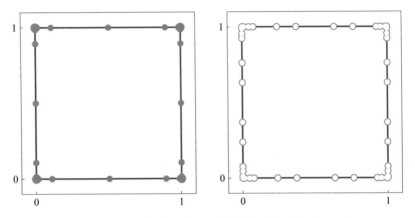

Fig. 5.175 Left: the knot locations on ∂S. Right: the collocation points.

The 32 knots determine 44 functions $b_{i,j}$ for each component of $\tilde{\psi}$, generating a total of 88 basis functions. The graphs of the $b_{i,j}$ in parametric form are shown in Fig. 5.176. Since the visual aspect of the functions constructed in the vicinity of the corners is not very helpful, in Fig. 5.177 we have redrawn the restrictions of the first four to the interval $0 \le t \le 0.017$, indicating the three spline knot locations and the four collocation points on them by black and white circular markers, respectively.

To compute $\tilde{\psi}$, we must choose an appropriate number of collocation points. The set of 64 unequally spaced values of the parameter t at our selection is

$\{0.001007, 0.002014, 0.006725, 0.01043, 0.04506, 0.07599, 0.2379,$
$0.3690, 0.6310, 0.7621, 0.9240, 0.9549, 0.9896, 0.9933, 0.9980,$
$0.9990, 1.001, 1.002, 1.007, 1.010, 1.045, 1.076, 1.238, 1.369, 1.631,$
$1.762, 1.924, 1.955, 1.990, 1.993, 1.998, 1.999, 2.001, 2.002, 2.007,$
$2.010, 2.045, 2.076, 2.238, 2.369, 2.631, 2.762, 2.924, 2.955, 2.990,$
$2.993, 2.998, 2.999, 3.001, 3.002, 3.007, 3.010, 3.045, 3.076, 3.238,$
$3.369, 3.631, 3.762, 3.924, 3.955, 3.990, 3.993, 3.998, 3.999\}.$

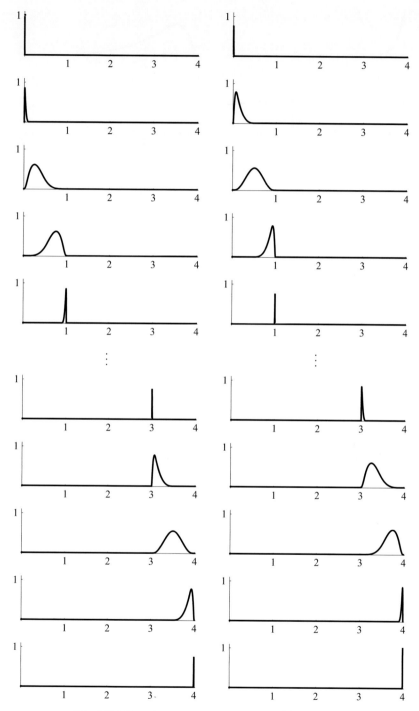

Fig. 5.176 The 44 B-spline basis functions $b_{i,j}[t]$, $0 \leq t \leq 4$.

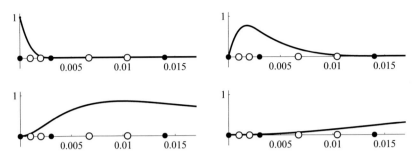

Fig. 5.177 The first four B-spline basis functions $b_{i,j}[t]$, $0 \leq t \leq 0.017$.

These values are marked on the graph on the right in Fig. 5.175. As in the case of the knots, only 32 of the 64 collocation points are visible, the remaining 32 being hidden under the four white corner markers. However, a few of the latter can be seen in Fig. 5.177; they correspond to the parametric locations

$$t = 0.001007, 0.002014, 0.006725, 0.01043.$$

The same 64 collocation points are used for both components of $\tilde{\psi}$, yielding a system of 128 constraining equations that enables us to compute the coefficients $c_{\alpha,i,j}$ in (5.68). The 88×128 matrix of this overdetermined system has a condition number of 281.

5.17.4 Solution

Approximate density. The two components of $\tilde{\psi}[x]$ are shown in Figs. 5.178 and 5.179. These graphs strongly suggest that the density $\psi[x[t]]$ is discontinuous at the primary knot locations $t = 0, 1, 2, 3, 4$. Additional computation shows that when supplementary spline knots and collocation points are placed closer to the corners, the spike in $\tilde{\psi}[x[t]]$ at these locations increases in height, which supports the conjecture that ψ is singular at the four corners.

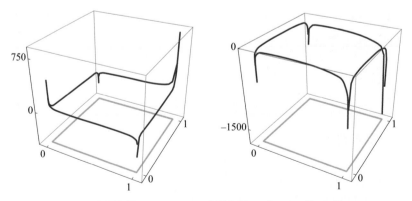

Fig. 5.178 The components of $\tilde{\psi}[x]$ (Cartesian coordinates).

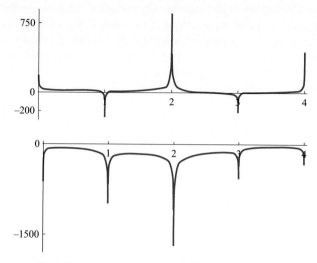

Fig. 5.179 The components of $\tilde{\psi}[x[t]]$ (parametric form).

Error analysis. A direct comparison between $\tilde{\psi}$ and ψ is not possible because the latter is unknown. However, we can do this indirectly by using the fact that u is computed from ψ by means of (5.67). The approximate density $\tilde{\psi}$ replaced in the restriction of this formula to the boundary produces an approximation $\tilde{u}$ to u on ∂S. The graphs of the components of both u (blue line) and $\tilde{u}$ (magenta line) for $x \in \partial S$ are shown in Fig. 5.180.

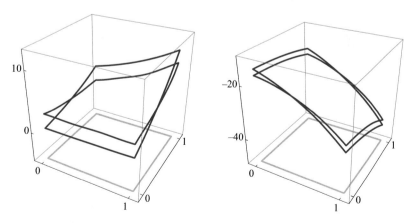

Fig. 5.180 The components of $u[x]$ (blue line) and $\tilde{u}[x]$ (magenta line), $x \in \partial S$.

These graphs tell us that the approximate and exact densities are in at least partial agreement for this example. However, the approximate density is not very accurate. Greater accuracy can, of course, be achieved by means of additional spline knots and collocation points that are even more densely packed around the corner locations. But this procedure becomes increasingly ill behaved. Here, we have used t values within 0.001 of a corner. A computation attempting to reduce this distance to 0.0001 requires a floating-point accuracy of over 70 digits, which very rapidly makes the problem unsolvable. Consequently, we do not recommend this method for problems involving boundaries with corners.

5.18 Remarks. (i) As illustrated in Sect. 5.16, the classical indirect method is computationally acceptable for smooth boundary curves.

(ii) The set of coefficients $\{c_{1,i,j}, c_{2,i,j}\}$ in (5.68), computed with the B-spline basis shown in Fig. 5.176, is

$$\{\{14., 16.2, 16.2, 19.9, 22.4, 54.6, -40.6, -10., 9.36, 33.7, 78.1, 178.,$$
$$3.09, 25.9, 12.1, 5.73, 2.75, 11.1, -17.1, -3.28, 2.4, 7.27, 16.7, 38.7\},$$
$$\{-86.8, -57.4, -56.7, -77.1, -135., -268., -205., -169., -159.,$$
$$-175., -219., -286., -470., -264., -167., -124., -117.,$$
$$-156., -79.3, -51.2, -35., -29.1, -33.4, -42.6\}\}.$$

5.18 Robin Problem in An Ellipse: Classical Indirect Method

5.18.1 Summary

This example illustrates the application of the classical indirect method to a Neumann problem in a domain with a smooth boundary. We compute the boundary condition from a known test solution, then use the collocation method with a piecewise cubic spline on equally spaced knots and collocation points to approximate the exact density. Since the latter is unknown, the results are validated through an indirect procedure. We also make a comparison between the classical indirect method for this problem and for the Neumann problem discussed in Sect. 5.16.

5.18.2 Problem Statement

Domain boundary. ∂S is the ellipse parameterized by

$$x1[t] = 1 + \mathrm{Cos}[\pi t], \quad x2[t] = \tfrac{1}{2}\mathrm{Sin}[\pi t] + \tfrac{1}{2}, \quad 0 \le t \le 2. \tag{5.69}$$

Its graph can be seen in Fig. 5.184.

Governing equations. The displacement u is the solution of the boundary value problem

$$Z_x \diamond u[x] = 0, \quad x \in S,$$
$$(\mathrm{T}u)[x] + \sigma u[x] = \mathscr{K}[x], \quad x \in \partial S.$$

Test solution. The two components of the test solution u given by (5.3) are graphed in Fig. 5.181.

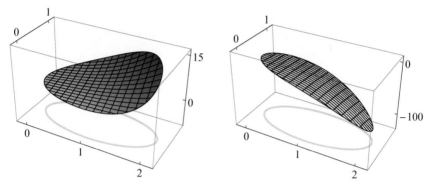

Fig. 5.181 The components of $u[x]$, $x \in S$.

Boundary data function. The function $\mathscr{K}$ is given by

$$(\mathrm{T}u)[x] + \sigma u[x] = \mathscr{K}[x], \quad x \in \partial S,$$

where, for computational purposes, we have chosen the particular 2×2 matrix

$$\sigma = \begin{pmatrix} 2 & 0 \\ 0 & 2 \end{pmatrix}.$$

The graphs of the two components of $\mathscr{K}$ are displayed in Figs. 5.182 and 5.183, in Cartesian coordinates and parametric form, respectively.

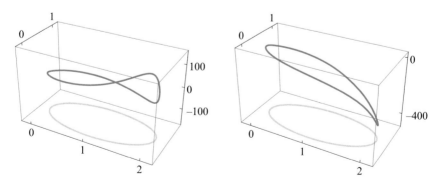

Fig. 5.182 The components of $\mathscr{K}[x]$ (Cartesian coordinates).

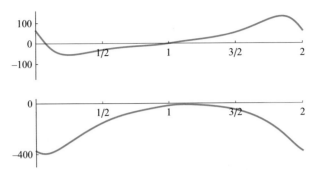

Fig. 5.183 The components of $\mathscr{K}[x[t]]$ (parametric form).

5.18.3 Solution Procedure

Method. We apply the classical indirect method, which reduces the problem to the boundary integral equation

$$\left(W_0^* + \tfrac{1}{2}\right)\varphi + V_0(\sigma\varphi) = \mathscr{K} \quad \text{on } \partial S,$$

coded as

$$\oint_{\Gamma_{\text{CPV}[x]}} (T_x \diamond D[x,y]) \circ \varphi[y]\, d\Gamma_y + \tfrac{1}{2}\,\varphi[x] + \oint_{\Gamma_{\text{Weak}[x]}} D[x,y] \circ (\sigma\varphi[y])\, d\Gamma_y = \mathscr{K}[x].$$

$$(5.70)$$

This equation is solved numerically to obtain an approximate density $\tilde{\varphi}$. In turn, $\tilde{\varphi}$ is used in the representation

$$u[x] = \oint_{\Gamma} D[x,y] \circ \varphi[y]\, d\Gamma_y, \quad x \in S \tag{5.71}$$

to generate an approximate solution $\tilde{u}$ in S.

Density function. The density φ does not represent either u or Tu on ∂S, which makes validation of the results somewhat more difficult.

Numerical approximation. We compute $\tilde{\varphi}$ by the collocation method with a B-spline basis of elements $b_{i,j}$. Then the approximate density is sought in the form

$$\tilde{\varphi}[x[t]] = \begin{pmatrix} \sum_i \sum_j c_{1,i,j} b_{i,j}[t] \\ \sum_i \sum_j c_{2,i,j} b_{i,j}[t] \end{pmatrix}, \tag{5.72}$$

where the coefficients $c_{\alpha,i,j}$ are determined by substituting (5.72) in (5.70).

   The construction of a spline basis requires that the smoothness of the spline at the knots must be the same as, or lower than, that of the function being approximated. Since the parameterization (5.69) is smooth, we expect φ to be equally smooth on ∂S, so we may construct an approximate density $\tilde{\varphi}$ with any degree of smoothness. Therefore, to ensure maximum smoothness, we choose a piecewise cubic spline that is twice continuously differentiable at the knots and has been modified to enforce periodicity and continuity at the knots corresponding to $t = 0, 2$. In terms of the parameterization (5.69), the knots and the smoothness at their locations are specified as the set

$$\left\{0, \frac{1}{6}, \frac{1}{3}, \frac{1}{2}, \frac{2}{3}, \frac{5}{6}, 1, \frac{7}{6}, \frac{4}{3}, \frac{3}{2}, \frac{5}{3}, \frac{11}{6}, 2\right\}.$$

The knots are marked on the graph on the left in Fig. 5.184.

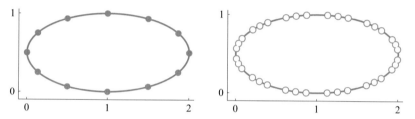

Fig. 5.184 Left: the knot locations on ∂S. Right: the collocation points.

   The 12 knots determine 12 functions $b_{i,j}$ for each component of $\tilde{\psi}$, generating a total of 24 basis functions. The graphs of the $b_{i,j}$ in parametric form are shown in Fig. 5.185. The first, second, and last basis functions have been modified to ensure periodic continuity at $t = 0, 2$.

   To compute $\tilde{\varphi}$, we must choose an appropriate number of collocation points. Owing to the smoothness of ∂S, these points can be placed almost anywhere. However, computational experiments (not shown here) indicate that the typical choice for these points generates a singular coefficient matrix. Consequently, to help with the computational stability of the method, we have augmented our original set of centrally placed locations with additional ones, which gives rise to an overdetermined system. The set of 36 equally spaced values of the parameter t at our selection is

$$\left\{\frac{1}{24}, \frac{1}{12}, \frac{1}{8}, \frac{5}{24}, \frac{1}{4}, \frac{7}{24}, \frac{3}{8}, \frac{5}{12}, \frac{11}{24}, \frac{13}{24}, \frac{7}{12}, \frac{5}{8}, \frac{17}{24}, \frac{3}{4}, \frac{19}{24}, \frac{7}{8}, \frac{11}{12}, \frac{23}{24},\right.$$
$$\left.\frac{25}{24}, \frac{13}{12}, \frac{9}{8}, \frac{29}{24}, \frac{5}{4}, \frac{31}{24}, \frac{11}{8}, \frac{17}{12}, \frac{35}{24}, \frac{37}{24}, \frac{19}{12}, \frac{13}{8}, \frac{41}{24}, \frac{7}{4}, \frac{43}{24}, \frac{15}{8}, \frac{23}{12}, \frac{47}{24}\right\}.$$

These values are marked on the graph on the right in Fig. 5.184.

   The same 36 collocation points are used for both components of $\tilde{\psi}$, yielding a system of 72 constraining equations that enables us to compute the coefficients $c_{\alpha,i,j}$ in (5.72). The 72×24 matrix of the overdetermined system has a condition number of 8.

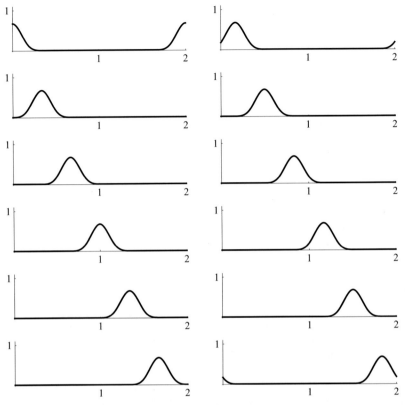

Fig. 5.185 The 12 B-spline basis functions $b_{i,j}[t]$, $0 \le t \le 2$.

5.18.4 Solution

Approximate density. The components of $\tilde{\varphi}$ are graphed in Figs. 5.186 and 5.187.

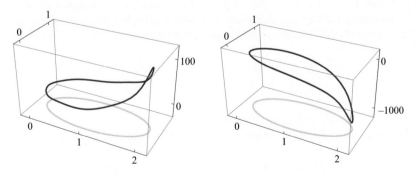

Fig. 5.186 The components of $\tilde{\varphi}[x]$ (Cartesian coordinates).

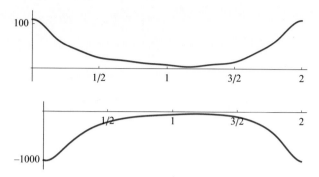

Fig. 5.187 The components of $\tilde{\varphi}[x[t]]$ (parametric form).

Approximate solution. We use $\tilde{\varphi}$ in (5.71) to compute an approximation $\tilde{u}$ to the exact solution u in S. The graphs of the two components of $\tilde{u}$ are shown in Fig. 5.188. There is virtually no visual distinction between $\tilde{u}$ in these graphs and u in Fig. 5.181.

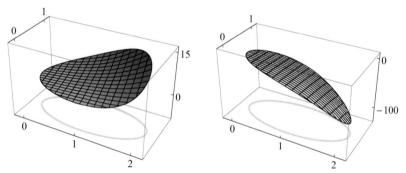

Fig. 5.188 The components of $\tilde{u}[x]$, $x \in S$.

Error analysis. A direct comparison between $\tilde{\varphi}$ and φ is not possible because the latter is unknown. However, we can do this indirectly by using the fact that u is computed from φ by means of (5.71). The approximate density $\tilde{\psi}$ replaced in the restriction of this formula to the boundary produces an approximation $\tilde{u}$ to u on ∂S. The difference $\tilde{u}[x] - u[x]$, $x \in \partial S$, relative to the maximum absolute value of u on ∂S is graphed in Figs. 5.189 and 5.190. The latter shows that the relative error in the two components of $\tilde{u}$ on the boundary is less than 1%, which confirms the validity of our computation.

Another procedure for illustrating the validity of our method is to compare the result for the Robin problem obtained above with that for the Neumann problem in Sect. 5.16 since in both cases the test solution u in S has the same representation; that is, equations (5.63) and (5.71) are essentially the same and generate the same function u, which means that the exact densities φ and ψ must coincide.

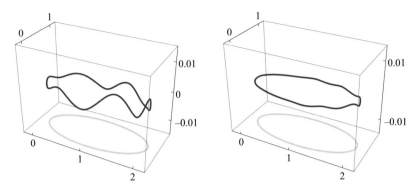

Fig. 5.189 The components of the relative error in $\tilde{u}[x]$, $x \in \partial S$ (Cartesian coordinates).

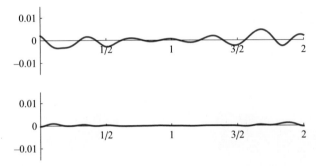

Fig. 5.190 The components of the relative error in $\tilde{u}[x[t]]$, $0 \leq t \leq 2$ (parametric form).

However, their approximations $\tilde{\varphi}$ and $\tilde{\psi}$ do not coincide because their computational constructions are different, even though they are based on boundary data functions $\mathcal{Q}$ and $\mathcal{K}$ in (5.64) and (5.72) that are both computed from the test solution (5.3). The difference $\varphi - \psi$ relative to the maximum absolute value of ψ on ∂S is graphed in Fig. 5.191. These graphs show that the relative size of the difference is less than 0.5%, which validates the results of the classical indirect method for both the Neumann and Robin boundary value problems.

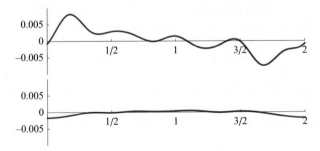

Fig. 5.191 The components of the relative difference $\varphi[x[t]] - \psi[x[t]]$.

5.19 Remarks. (i) We mention that the density φ in this problem suffers from the same computational issues as the density ψ in Sect. 5.17 as they are both used in the same integral representation of the exact solution u. Consequently, if the boundary has corners, we would encounter the same computational difficulties when constructing $\tilde{\varphi}$ as we did for $\tilde{\psi}$.

(ii) The set of coefficients $\{c_{1,i,j}, c_{2,i,j}\}$ in (5.72), computed with the B-spline basis shown in Fig. 5.185, is

$$\{\{133., 61.4, 42.8, 20.4, 19.5, 10.5, 9.74, 0.881, 11.4,$$
$$10.3, 37.9, 61.5\}, \{-1200., -735., -377., -208., -116.,$$
$$-81.1, -62.7, -37.1, -53.9, -88., -220., -496.\}\}.$$

5.19 Mixed Boundary Conditions: Piecewise Cubic Spline

5.19.1 Summary

This example illustrates the application of the direct method to a mixed problem in a semicircular domain with Dirichlet conditions prescribed on the diameter and Neumann conditions on the arc. Both data functions are computed from a known test solution. We use the collocation method with a piecewise cubic spline on equally spaced knots and collocation points to construct an approximation of the original solution from the exact boundary data functions. The results are validated through a comparison between the exact and approximate solutions.

5.19.2 Problem Statement

Domain boundary. ∂S is a closed curve consisting of a straight line segment ∂S_1 and the upper half circle ∂S_2 that has this segment as its diameter. It is parameterized by

$$x1[t] = \begin{cases} 2t, & 0 \leq t \leq 1, \\ 1 + \mathrm{Cos}[\pi(t-1)], & 1 < t \leq 2, \end{cases}$$

$$x2[t] = \begin{cases} 0, & 0 \leq t \leq 1, \\ \mathrm{Sin}[\pi(t-1)], & 1 < t \leq 2, \end{cases}$$

(5.73)

and its graph can be seen in Fig. 5.201.

Governing equations. The displacement u is the solution of the boundary value problem

$$Z_x \diamond u[x] = 0, \quad x \in S,$$
$$u[x] = \mathscr{P}[x], \quad x \in \partial S_1,$$
$$(\mathrm{Tu})[x] = \mathscr{Q}[x], \quad x \in \partial S_2.$$

Test solution. The two components of the test solution u given by (5.3) are graphed in Fig. 5.192.

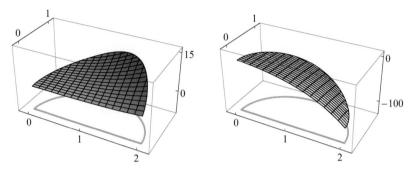

Fig. 5.192 The components of $u[x]$, $x \in S$.

Boundary data functions. The function $\mathscr{P}$ is given by (5.3) on ∂S_1; its components are graphed in Fig. 5.193 and 5.194, in Cartesian coordinates and parametric form, respectively.

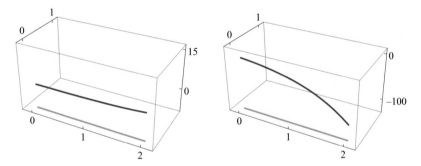

Fig. 5.193 The components of $\mathscr{P}[x]$, $x \in S_1$ (Cartesian coordinates).

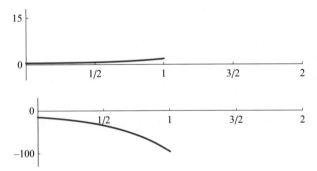

Fig. 5.194 The components of $\mathscr{P}[x[t]]$, $0 \leq t \leq 1$ (parametric form).

The function $\mathcal{Q}$ is given by (5.4) on ∂S_2; its components are displayed in Figs. 5.195 and 5.196.

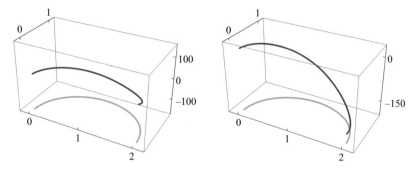

Fig. 5.195 The components of $\mathcal{Q}[x]$, $x \in \partial S_2$ (Cartesian coordinates).

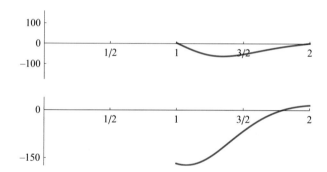

Fig. 5.196 The components of $\mathcal{Q}[x[t]]$, $1 < t < 2$ (parametric form).

5.19.3 Solution Procedure

Method. We apply the direct method, which reduces the problem to the boundary integral equations

$$V_0(\varphi) = \left(W_0 + \tfrac{1}{2}I\right)\mathscr{P} \quad \text{on } \partial S_1,$$

codes as

$$\oint_{\Gamma_{\text{Weak}[x]}} D[x,y] \circ \varphi[y]\, d\Gamma_y = \oint_{\Gamma_{\text{CPV}[x]}} P[x,y] \circ \mathscr{P}[y]\, d\Gamma_y + \tfrac{1}{2}\,\mathscr{P}[x], \quad x \in \partial S_1,$$

$$(5.74)$$

where $\varphi[x] = (\mathrm{T}u)[x]$, and

$$\left(W_0 + \tfrac{1}{2}I\right)\varphi = V_0 \mathcal{Q} \quad \text{on } \partial S_2,$$

coded as

$$\oint_{\Gamma_{\mathrm{CPV}[x]}} P[x,y] \circ \varphi[y]\,d\Gamma_y + \tfrac{1}{2}\,\varphi[x] = \oint_{\Gamma_{\mathrm{Weak}[x]}} D[x,y] \circ \mathcal{Q}[y]\,d\Gamma_y, \quad x \in \partial S_2,$$

(5.75)

where $\varphi[x] = u[x]$, $x \in \partial S_2$. These equations are solved numerically to obtain an approximate density $\tilde{\varphi}$. In turn, $\tilde{\varphi}$ is used in the representation formula

$$u[x] = \oint_{\partial S_1} D[x,y] \circ \varphi[y]\,d\Gamma_y + \oint_{\partial S_2} D[x,y] \circ \mathcal{Q}[y]\,d\Gamma_y$$

$$- \oint_{\partial S_1} P[x,y] \circ \mathscr{P}[y]\,d\Gamma_y - \oint_{\partial S_2} P[x,y] \circ \varphi[y]\,d\Gamma_y, \quad x \in S,$$

(5.76)

to generate an approximate solution $\tilde{u}$ in the domain S.

Density function. The exact density φ on ∂S_2 is the restriction to ∂S_2 of the test solution u; its components are graphed in Figs. 5.197 and 5.198 in Cartesian coordinates and parametric form, respectively.

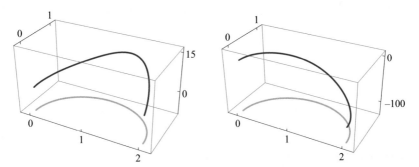

Fig. 5.197 The components of $\varphi[x]$, $x \in \partial S_2$ (Cartesian coordinates).

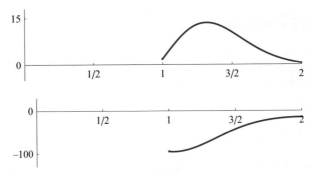

Fig. 5.198 The components of $\varphi[x[t]]$, $1 < t < 2$ (parametric form).

The exact density φ on ∂S_1 is the restriction to ∂S_1 of the boundary stress vector Tu computed from the test solution and given by (5.4); its components are graphed in Figs. 5.199 and 5.200.

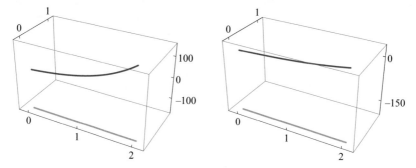

Fig. 5.199 The components of $\varphi[x]$, $x \in \partial S_1$ (Cartesian coordinates).

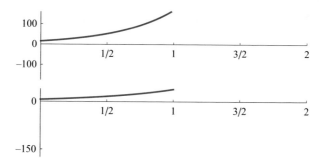

Fig. 5.200 The components of $\varphi[x[t]]$, $0 < t < 1$ (parametric form).

Numerical approximation. We compute $\tilde{\varphi}$ by the collocation method with a B-spline basis of elements $b_{i,j}$. Then the approximate density is sought in the form

$$\tilde{\varphi}[x[t]] = \begin{pmatrix} \sum_i \sum_j c_{1,i,j} b_{i,j}[t] \\ \sum_i \sum_j c_{2,i,j} b_{i,j}[t] \end{pmatrix}, \tag{5.77}$$

where the numerical coefficients $c_{\alpha,i,j}$ are determined by substituting (5.77) in (5.74) and (5.75).

Since φ is discontinuous at the two corners of ∂S because of the boundary condition switch between $\mathscr{P}$ and $\mathscr{Q}$ at those points, the $b_{i,j}$ need to exhibit the same feature. In this case, we choose a piecewise cubic spline that is twice continuously differentiable at the secondary knots and discontinuous at the primary knot locations $t = 0, 1, 2$. In terms of the parameterization (5.73), the knots and the smoothness at their locations are specified as the set

$$\left\{ 0, 0, 0, 0, \tfrac{1}{6}, \tfrac{1}{3}, \tfrac{1}{2}, \tfrac{2}{3}, \tfrac{5}{6}, 1, 1, 1, 1, \tfrac{7}{6}, \tfrac{4}{3}, \tfrac{3}{2}, \tfrac{5}{3}, \tfrac{11}{6}, 2, 2, 2, 2 \right\}.$$

The knots are marked on the graph on the left in Fig. 5.201.

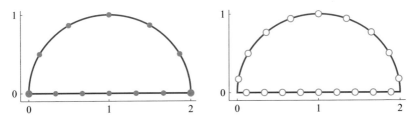

Fig. 5.201 Left: the knot locations on ∂S. Right: the collocation points.

The 12 knots determine 18 functions $b_{i,j}$ for each component of $\tilde{\psi}$, generating a total of 36 basis functions. The first nine have support only on ∂S_1, and the remaining nine have support only on ∂S_2. The graphs of the $b_{i,j}$ in parametric form are shown in Fig. 5.202.

The computed density $\tilde{\phi}[x[t]]$ that approximates $(\mathrm{Tu})[x[t]]$ on $0 < t < 1$ is expected to 'join' the boundary condition function $\mathcal{Q}[x[t]]$ (defined on $1 < t < 2$) with jump discontinuities at $t = 0, 1, 2$ because the parameterized boundary contour $\partial S = \partial S_1 \cup \partial S_2$ has corners at these locations. Therefore, the first nine basis functions $b_{i,j}$ for ∂S_1 in Fig. 5.202 are acceptable in their current form. On the other hand, the computed density $\tilde{\phi}[x[t]]$ that approximates $u[x[t]]$ on $1 < t < 2$ is expected to join the boundary condition function $\mathcal{P}[x[t]]$ (defined on $0 < t < 1$) without a jump at those locations, which suggests that we should impose an additional restriction on the second nine basis functions to guarantee continuity at $t = 0, 1, 2$. Specifically, we choose the coefficients at positions 9 and 18 in Fig. 5.202 to be

$$\tilde{\phi}[x[1]] = u[x[1]] = \{1.85013, -95.3664\},$$
$$\tilde{\phi}[x[2]] = u[x[0]] = \{0.476909, -14.8369\}.$$

This means that the boundary element method will compute only seven of the nine basis functions for each of the two components on ∂S_2.

To compute $\tilde{\phi}$, we must choose an appropriate number of collocation points, which can be placed anywhere on the boundary except at the corners, where $\tilde{\phi}$ is not defined. The set of values of the parameter t at our selection is

$$\left\{ \frac{1}{18}, \frac{1}{6}, \frac{5}{18}, \frac{7}{18}, \frac{1}{2}, \frac{11}{18}, \frac{13}{18}, \frac{5}{6}, \frac{17}{18}, \frac{19}{18}, \frac{7}{6}, \frac{23}{18}, \frac{25}{18}, \frac{3}{2}, \frac{29}{18}, \frac{31}{18}, \frac{11}{6}, \frac{35}{18} \right\}.$$

These values are marked on the graph on the right in Fig. 5.201.

The first nine collocation points lie on ∂S_1 and are used to determine the coefficients of the first nine basis functions for $x \in \partial S_1$; the other nine lie on ∂S_2 and are used to determine the coefficients of the last seven basis functions for $x \in \partial S_2$. The same 18 collocation points are used for both components of $\tilde{\phi}$, yielding a system of 36 constraining equations that enables us to compute the numerical coefficients $c_{\alpha,i,j}$ in (5.77). The overdetermined 36×32 matrix of this system has a condition number of 428.

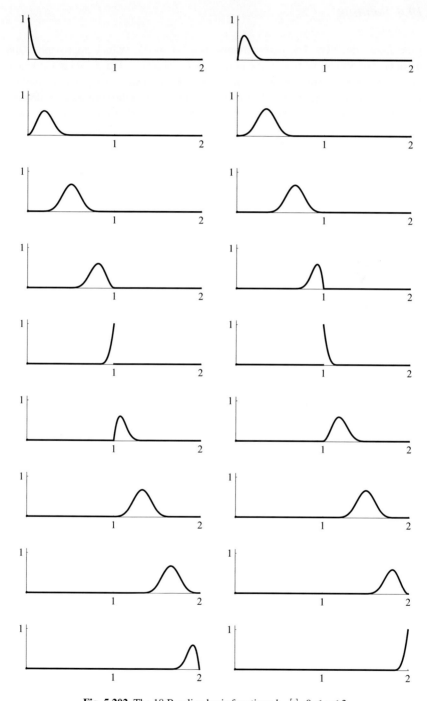

Fig. 5.202 The 18 B-spline basis functions $b_{i,j}[t]$, $0 \leq t \leq 2$.

5.19.4 Solution

Approximate density. The two components of $\tilde{\varphi}$ on ∂S_1, which approximate the components of Tu on that part of the boundary, are shown in Fig. 5.203 in Cartesian coordinates. Their parametric form for $0 < t < 1$, and the primary and secondary knots, can be seen in Fig. 5.204. These graphs indicate that the approximate density $\tilde{\varphi}$ on ∂S_1 agrees very well with the exact one $\varphi = $ Tu displayed in Figs. 5.199 and 5.200.

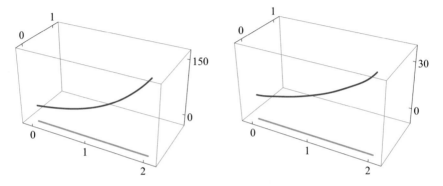

Fig. 5.203 The components of $\tilde{\varphi}[x]$, $x \in \partial S_1$ (Cartesian coordinates).

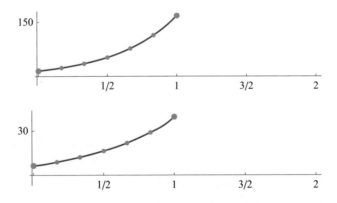

Fig. 5.204 The components of $\tilde{\varphi}[x[t]]$, $0 < t < 1$ (parametric form).

The components of φ on ∂S_2, which approximate those of u there, are graphed in Fig. 5.205. (We recall that two coefficients of two basis functions for both components have been modified to prevent jump discontinuities.) Their parametric form for $1 < t < 2$, and the primary and secondary knots, can be seen in Fig. 5.206. These graphs indicate that the approximate density $\tilde{\varphi}$ on ∂S_2 agrees very well with the exact one $\varphi[x] = u[x]$ $x \in \partial S_2$, displayed in Figs. 5.197 and 5.198.

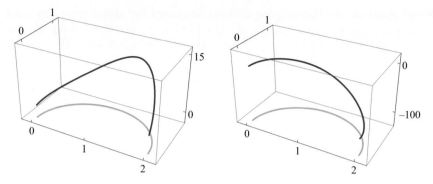

Fig. 5.205 The components of $\tilde{\varphi}[x]$, $x \in \partial S_2$ (Cartesian coordinates).

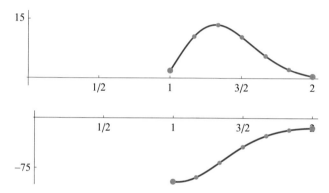

Fig. 5.206 The components of $\tilde{\varphi}[x[t]]$, $1 < t < 2$ (parametric form).

Approximate solution. We use $\tilde{\varphi}$, $\mathscr{P}$, and $\mathscr{Q}$ in (5.76) to compute an approximation $\tilde{u}$ to the exact solution u in S. The graphs of the two components of $\tilde{u}$ are shown in Fig. 5.207.

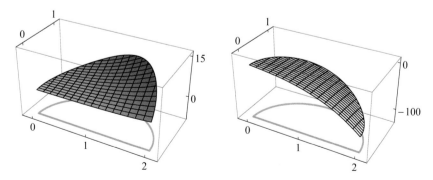

Fig. 5.207 The components of $\tilde{u}[x]$, $x \in S$.

Error analysis. To visualize the difference between the approximate and exact densities, in Fig. 5.208 we graphed $\tilde{\varphi} - \varphi$ relative to the maximum absolute value of φ on ∂S. The same graphs in parametric form in Fig. 5.209, where the intervals $0 < t < 1$ and $1 < t < 2$ correspond to ∂S_1 and ∂S_2, respectively, show that the relative error in $\tilde{\varphi}$ on the entire boundary with the exception of the corners is less than 1%.

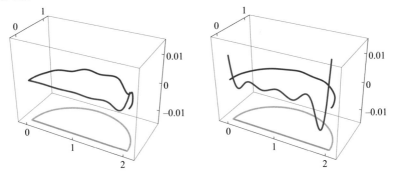

Fig. 5.208 The components of the relative error in $\tilde{\varphi}[x]$ (Cartesian coordinates).

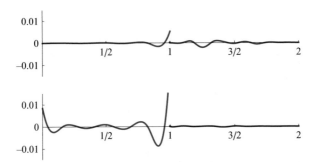

Fig. 5.209 The components of the relative error in $\tilde{\varphi}[x[t]]$ (parametric form).

5.20 Remarks. (i) The direct method for the Dirichlet, Neumann, and mixed problems appears to be well conditioned as well as computationally accurate. In general, comparing this method with the classic indirect method illustrated in the preceding sections, we conclude that the former has been, from a computational viewpoint, consistently superior to the latter.

(ii) The set of coefficients $\{c_{1,i,j}, c_{2,i,j}\}$ in (5.77), computed with the B-spline basis shown in Fig. 5.202, is

$$\{\{\{15.1, 17.4, 22.9, 34.3, 51., 75.1, 111., 145., 168.\},$$
$$\{1.85, 5.02, 11.7, 14.6, 10.6, 5.31, 1.92, 0.877, 0.476\}\},$$
$$\{\{6.41, 6.64, 8.86, 11.8, 16.2, 21.3, 28.9, 33.9, 39.5\},$$
$$\{-95.3, -96.8, -91.6, -66.9, -41.5, -25., -16.8, -15., -14.8\}\}\}.$$

References

1. Barber, J.R.: Elasticity, 2nd edn. Kluwer, London (2002)
2. Chudinovich, I., Constanda, C.: Variational and Potential Methods in the Theory of Bending of Plates with Transverse Shear Deformation. Chapman & Hall/CRC, Boca Raton (2000)
3. Chudinovich, I., Constanda, C.: Variational and Potential Methods for a Class of Linear Hyperbolic Evolutionary Processes. Springer, London (2005)
4. Chudinovich, I., Constanda, C., Doty, D., Hamill, W., Pomeranz, S.: On a boundary value problem for the plane deformation of a thin plate on an elastic foundation. In: Proceedings of the Thirteenth International Symposium on Methods of Discrete Singularities in Problems of Mathematical Physics, Kharkov-Kerson, pp. 358–361 (2007)
5. Chudinovich, I., Constanda, C., Doty, D., Hamill, W., Pomeranz, S.: The Dirichlet problem for the plane deformation of a thin plate on an elastic foundation. In: Constanda, C., Potapenko, S. (eds.) Integral Methods in Science and Engineering: Techniques and Applications, pp. 83–88. Birkhäuser, Boston (2008)
6. Constanda, C.: Direct and Indirect Boundary Integral Equation Methods. Chapman & Hall/CRC, Boca Raton (2000)
7. Constanda, C.: Mathematical Methods for Elastic Plates. Springer, New York (2014)
8. Evans, L.C.: Partial Differential Equations. American Mathematical Society, Providence (1998)
9. Hetnarski, R.B., Ignaczak, J.: Mathematical Theory of Elasticity. Taylor & Francis, London (2004)
10. Kreyszig, E.: Advanced Engineering Mathematics. Wiley, New York (1972)
11. Lopez, R.J.: Advanced Engineering Mathematics. Addison Wesley, Boston (2001)
12. Marsden, J.M., Hughes, T.J.: Mathematical Foundations of Elasticity. Dover, New York (1994)
13. McOwen, R.C.: Partial Differential Equations: Methods and Application, 2nd edn. Prentice Hall River, Upper Saddle (2003)
14. Mikhlin, S.G.: Integral Equations and Their Application in Mechanics, Mathematical Physics and Technology. Pergamon Press, Oxford (1957)
15. Nédélec, J.-C.: Acoustic and Electromagnetic Equations. Springer, Heidelberg (2000)
16. Thomson, G.R., Constanda, C.: Stationary Oscillations of Elastic Plates. Birkhäuser, Boston (2011)
17. Vladimirov, V.S.: Equations of Mathematical Physics. Marcel Dekker, New York (1971)
18. Winkler, E.: Die Lehre von der Elastizität und Festigkeit. Dominicus, Prague (1867)

© Springer International Publishing Switzerland 2016
C. Constanda et al., *Boundary Integral Equation Methods and Numerical Solutions*,
Developments in Mathematics 35, DOI 10.1007/978-3-319-26309-0

Index

© Springer International Publishing Switzerland 2016
C. Constanda et al., *Boundary Integral Equation Methods and Numerical Solutions*,
Developments in Mathematics 35, DOI 10.1007/978-3-319-26309-0

Printed in the United States
By Bookmasters